OTTO HAHN AND THE RISE OF NUCLEAR PHYSICS

OTTO HAHN 1935

OTTO HAHN AND THE RISE OF NUCLEAR PHYSICS

Edited by

WILLIAM R. SHEA

McGill University, Montreal, Canada

D. REIDEL PUBLISHING COMPANY

A MEMBER OF THE KLUWER ACADEMIC PUBLISHERS GROUP

DORDRECHT / BOSTON / LANCASTER

Library of Congress Cataloging in Publication Data

Main entry under title:

Otto Hahn and the rise of nuclear physics.

(The University of Western Ontario series in philosophy of
science ; v. 22)
Includes index.
Contents: Introduction / William R. Shea — The nuclear electron
hypothesis / Roger H. Stuewer — The evolution of matter / Robert H.
Kargon — [etc.]
1. Nuclear physics—History—Addresses, essays, lectures.
2. Hahn, Otto, 1879–1968—Addresses, essays, lectures. 3. Physicists—
Germany—Biography—Addresses, essays, lectures. I. Shea, William R.
II. Series.
QC773.087   1983   539.7′092′4   83–10967
ISBN 90–277–1584–X

Published by D. Reidel Publishing Company,
P.O. Box 17, 3300 AA Dordrecht, Holland.

Sold and distributed in the U.S.A. and Canada
by Kluwer Academic Publishers,
190 Old Derby Street, Hingham, MA 02043, U.S.A.

In all other countries, sold and distributed
by Kluwer Academic Publishers Group,
P.O. Box 322, 3300 AH Dordrecht, Holland.

Printed in The Netherlands

TABLE OF CONTENTS

ACKNOWLEDGMENTS

The editor gratefully acknowledges the generous support of Vice Principals Walter Hitschfeld and Leo Yaffe of McGill University, of Dr. Philipp Schmidt-Schlegel, the Consul General of the Federal Republic of Germany in Montreal, and of Dr. Heinz Hugo Becker, the Director of the Goethe Institute in Montreal. He also wishes to express his warm thanks to the Social Sciences and Humanities Research Council of Canada and the Deutsche Forschunggesellschaft for making this volume possible.

INTRODUCTION: FROM RUTHERFORD TO HAHN

When Otto Hahn went abroad to work with Ernest Rutherford at McGill University in Montreal in the autumn of 1905 he was already twenty-six years old and had been a doctor of chemistry since 1901. The organic chemist Theodor Zincke had supervised his research at Marburg, and after Hahn returned from military service in 1902, he had offered him an appointment as his laboratory assistant, to give him sufficient experience to qualify for employment in one of Germany's large chemical firms. In 1904, the head of Kalle and Company mentioned to Zincke that they would like to hire a young chemist but that the candidate would have to be familiar with a foreign country and be fluent in a second language. Zincke suggested to Hahn that he go to England to learn English and to work under Sir William Ramsay, the famous discoverer of the inert gases, with whom Zincke entertained friendly relations. Hahn arrived at the Chemical Institute of University College in London in the autumn of 1904, and Ramsay immediately asked him whether he wanted to work with radium. When Hahn replied that he did not know anything about radium, Ramsay said that this would do no harm since it would enable him to approach the project with an open mind. He handed Hahn a 100-gram specimen of barium chloride and told him to use Marie Curie's method for extracting radium. She had found that radium is slightly more difficult to dissolve than barium, but when recrystallized it precipitates more abundantly, although always with barium. If the process is interrupted and then restarted, a noticeable enrichment of radium can be obtained after several repetitions. This technique, known as fractional crystallization, was the one that Hahn rapidly mastered.

Some time later, Ramsay gave Hahn and his fellow countryman Otto Sackur samples of actinium and emanium about which a controversy was raging in the scientific community. The first had been discovered by André Debierne in France and the second by Friedrich Giesel in Germany. Only a study of their radioactive properties could determine whether they were the same substance or two different elements. Hahn and his colleague wrapped the preparations in paper, inserted them into a cotton-filled glass tube, and blew air through the tube into another container where the radioactive air was measured with an electrometer. The pointer of the electrometer rose less

1

William R. Shea (ed.), Otto Hahn and the Rise of Nuclear Physics, 1–18.
Copyright © 1983 by D. Reidel Publishing Company.

and less as the emanation underwent radioactive decay, and it became motionless after about 30 seconds. Since this process was occurring very rapidly, Hahn and Sackur marked the position of the pointer on a scale with pencil marks. As a timing device they used a metronome that beat out intervals of approximately 1.3 seconds. This simple method enabled them to determine that the half-life of the emanations of actinium and emanium were the same. Although Giesel's measurements had been more precise than Debierne's, the name of actinium was retained since Debierne had made the discovery first.

Hahn now returned to his sample of barium chloride. He soon conjectured that the radium-enriched preparations must harbor another radioactive substance. The liquids resulting from fractional crystallization, which were supposed to contain radium only, produced two kinds of emanation. One was the long-lived emanation of radium, the other had a short life similar to the emanation produced by thorium. Hahn tried to separate this substance by adding some iron to the solutions that should have been free of radium, but to no avail. Later the reason for his failure became apparent. The element that emitted the thorium emanation was constantly replenished by the element believed to be radium. Hahn succeeded in enriching a preparation until it was more than 100,000 times as intensive in its radiation as the same quantity of thorium. He concluded that this was a new element, and he christened it radiothorium. In the light of the subsequent discovery of isotopes, Hahn's "emanation of thorium" was identified as the noble gas radon (radon isotope 220) and his "radiothorium" as a new isotope of thorium, the isotope with mass number 228.

Ramsay was enthusiastic about Hahn's discovery and urged him to engage upon a career of detection of new radioactive elements. Ramsay often nursed sanguine expectations about new substances, sometimes to his discomfiture. When a colleague at the Chemical Institute, a certain professor Ogawa, thought that he had found a new element which resembled tin but was separable from it, Ramsay delivered a lecture at the Royal Society on the new element and gave it the name "niponium." Unfortunately, it eluded all further research. Hahn himself was almost a victim of Ramsay's leaps of faith. In order to precipitate his thorium preparations, Hahn occasionally used hydrogen sulfide, and he noticed one day that he had obtained a thin film of precipitation. He told this to Ramsay, who said at once, "That's a new stuff." He boldly inferred that the radiothorium had changed into something else, and he suggested sending a short communication to the Royal Society. But Hahn's experimental caution was as great as his experimental flair, and he asked Ramsay to wait until he had better results. The "new stuff" turned out to be caused by the dust that fell from the rusty iron of the roof of the laboratory.

Ramsay's willingness to welcome a plethora of new elements must be seen against the background of the chemical theory of the day. At a later date a satisfactory model of the atomic nucleus would show that only ninety-two natural elements were possible, but at this time a proliferation of elements could not be ruled out. Indeed, "new" elements were periodically announced. While he was in London, Ramsay introduced Hahn to a Professor Baskerville, who claimed to have separated thorium into three different elements, which he called berzelium, carolinium, and neothorium. In spite of an offer of a job at his institute in the United States, Hahn does not appear to have taken the professor's alleged discovery very seriously.

On the day of his arrival in London, Hahn found that skepticism could even extend to metal currency. He climbed to the top of a double-decker bus to see something of the city during the ride, and when he handed a golden sovereign to the conductor the man turned it over and returned it to him saying something that Hahn did not understand. Since he had no small change, he gave the gold piece back to the conductor, who again returned it with a few words that Hahn again found unintelligible. Fortunately another passenger noticed Hahn's baedeker and came up to explain in German that the conductors were not allowed to accept gold on the upper deck of the bus because circumstances made it impossible for them to check the genuineness of the coin by making it ring. He gave him a twopence and Hahn paid his fare.

Working conditions in Ramsay's institute were congenial but a far cry from those of a modern laboratory. When Hahn needed space for the electroscope with which he could measure the success of his attempts to enrich a sample with radiothorium, Ramsay provided him with a key that unlocked all the doors of University College and told him to find a suitable place. Hahn succeeded in securing a table in the preparation room of the Physical Institute for his electroscope, but he could not use it during the day when the room was needed for the preparation of laboratory experiments for the students. Hahn then got into the habit of returning to the Physical Institute after 9 p.m., and for several months he took his painstaking measurements in the still of the night when he could work undisturbed.

At the end of his two-year stay in London, Ramsay urged Hahn not to go into industry but to embark upon a university career, preferably in Berlin, the leading center of research in Germany. When Hahn replied that he knew no one in Berlin, Ramsay immediately wrote a glowing letter of reference to his friend Emil Fischer, the director of the Chemical Institute at the University of Berlin. He praised Hahn for his ingenuity and experimental skill in isolating radiothorium, and he warmly recommended him as "a nice fellow, modest, completely trustworthy, and very gifted."[1] Fischer was willing to find room for Hahn in his laboratory, but Hahn, who had no pressing needs for an immediate income, preferred to pursue his work in radioactivity with Ernest Rutherford, the leading authority in the field.

In his letter of inquiry to Rutherford, Hahn mentioned that he had found a new radioactive element, a claim that Rutherford immediately doubted. His valid reasons for feeling hesitant about Hahn's alleged discovery are discussed by Thaddeus Trenn in his essay in this volume. They throw considerable light on the state of physics in 1905. In spite of his reservations, Rutherford wrote back to say that he would be glad to welcome Hahn and that details could be arranged when he was in Montreal. Shortly after his arrival in September 1905 Hahn was able to persuade Rutherford that he had really found something new, not merely the short-lived thorium-C that Rutherford and Frederick Soddy had discovered. Rutherford later admitted to Hahn that he had been influenced by his friend Bertram B. Boltwood of Yale, who had rashly qualified Hahn's substance as "a mixture of thorium and stupidity."[2]

Boltwood was soon made to acknowledge that Hahn had discovered a genuine element, but he disputed the value of its half-life. Hahn's direct measurement of the emission indicated that radiothorium had a half-life of about two years. Boltwood had analyzed a number of thorium compounds that were on the market and had found that they were weak in radioactivity compared with minerals containing thorium. He reasoned that if the radiothorium, which is mainly responsible for the radiation coming from thorium, had a half-life of two years, its recurring formation should be detected with the electroscope within a matter of months. Since he did not find this to be the case, he concluded that Hahn had been slipshod in his experimental technique and that radiothorium must have a longer half-life. Rutherford, who revelled in vigorous debate, brought the two contestants together. Neither was able to persuade the other, but Hahn made the suggestion that their experimental results could be made to tally if one assumed an unknown radioactive element, with a long half-life, between thorium and radiothorium. Boltwood dismissed the suggestion as a mere hypothesis, but Hahn was to have the pleasure of being vindicated a year or so later when he investigated thorium compounds that had been prepared over a period of ten years. Hahn found that recently prepared salts showed the same amount of activity as minerals containing the same quantity of thorium. As the samples aged, their radioactivity declined. Preparations that were several years old displayed only half as much activity as freshly prepared salts. Preparations that were ten years or older, however, showed a higher activity but never as much as in the newly prepared salts. Hahn realized that his conjecture provided the answer to this problem. When thorium is extracted from a mineral, an unidentified nonradiating substance, the source of radiothorium, remains with the mineral, while radiothorium is extracted along with the thorium. Since the half-life of radiothorium is about two years, the radioactivity of the mixture drops appreciably. But this is counteracted by the production of radiothorium from the unidentified substance that is constantly being formed by thorium. As Hahn had assumed, if this middle substance, which he called mesothorium, has a longer half-life than radiothorium, the activity of a given preparation will decline until it reaches its lowest ebb and then will begin to increase again until an old thorium salt is almost as powerful as a fresh sample. Hahn rightly concluded that Boltwood must have been using samples that were about three years old when the changes in activity are so small that they are scarcely discernible.

The Macdonald Physics Laboratory at McGill, where Rutherford came to work in 1898, had been richly endowed by the eccentric millionaire and tobacco king Sir William Macdonald. In Rutherford's own words, it was "the best of its kind in the world,"[3] but at the turn of the century this meant little more than the availability of decent facilities. Instrumentation was still primitive and had to be constructed by the experimenters themselves. The beta-ray and gamma-ray electroscopes were made of large sheet metal cans on which tobacco or coffee cans were mounted. The insulation of the support for the two metal foils was sulfur, presumably because amber was not readily available in Canada. The evacuation of Rutherford's apparatus for his experiments with alpha particles was accomplished by means of an old-fashioned Töpler pump. This involved repeatedly raising and lowering a bulb containing mercury, a manual process that easily took the best part of a morning, with the result that the sample to be investigated had largely decayed before a sufficiently good vacuum was produced.

Hahn did not arrive in Montreal empty-handed. Sir William Ramsay had given him the radiothorium that he had separated and the actinium that he and Sackur had proved to be identical with Giesel's emanium. These gifts were by no means negligible, but radioactive preparations were not as expensive at the beginning of the century as they were to become during World War I. This was largely because Giesel, as a sideline, produced and sold radium salts at a very reasonable price. In 1903 a milligram of radium bromide could still be purchased for 5 dollars.

When Hahn came to Montreal the period of collaboration between Rutherford and Soddy, who had arrived at McGill in May 1900, had already ended with Soddy's return to England in 1903. This had been one of the most productive associations between two scientists in the history of physics. Rutherford knew little chemistry — a detailed knowledge of which was essential for the separation of the radioactive elements — but he was an unrivalled expert in the measurement of radioactivity by ionization methods and in the construction of the apparatus required. Soddy was a gifted chemist but had no experience with radioactivity. Both were enthusiastic about new experiments and fascinated by the possibilities of the atomic theory. They soon found that a very active substance, which they called thorium X, could be separated from thorium by a simple chemical operation. The thorium freed from thorium X lost its radioactivity in an exponential manner, the time to half activity being about four days. The two scientists showed that this behavior could be easily explained if the thorium recovered its lost activity by forming thorium X at a constant rate, while the thorium X so formed decayed in the exponential way characteristic of radioactive substances. In 1903 Rutherford succeeded in showing that the alpha rays that he had discovered five years earlier were deflected by both an electric and a magnetic field. The direction of the deflection showed that they were positively charged, and the ratio of the charge to the mass was found to be roughly one half of what it was for the hydrogen atom. Hence if the charge was equal to the charge on the electron, but positive, the mass must be twice that of the hydrogen atom; or, as Rutherford and Soddy conjectured as early as 1902, if the charge was double the electronic charge, then the mass must be four times that of the hydrogen atom, namely the mass of the helium atom. The second alternative is correct, but it was not until 1908 that Rutherford, working with Hans Geiger in Manchester, was able to settle the issue.

Rutherford then considered looking for an experimental proof of the formation of helium from radium. When Soddy returned to England it was with the understanding that he was to try to establish that radium is produced by uranium — still a moot question — while Rutherford focused on the formation of helium from radium. Upon his arrival in Ramsay's laboratory, however, Soddy began working on radium bromide and, with the aid of spectroscopy, he soon succeeded with Ramsay in showing that helium comes from radium. Rutherford courteously sent his own radium to Soddy to help him verify his discovery, but the matter weighed on him, and when Hahn came to McGill he found Rutherford as eloquent on Soddy's scientific merits as he was silent on his personal qualities.

As soon as Rutherford had satisfied himself that Hahn's radiothorium was a genuine element he became greatly interested in it as a source of alpha particles, and Hahn set out to determine the range of alpha particles from the radioactive precipitation as well as from radiothorium and thorium X. The range of these particles was determined by counting their impact on a zinc sulfide screen, and Hahn was soon able to show that the alpha particles coming from the active deposit of thorium were much faster than the fastest then known, which were those emitted by radium C. While working on the active precipitate of radiothorium he found that it did not only contain thorium A, with a half-life of 10.6 hours, and thorium B, with a half-life of 55 minutes, but another alpha-emitting substance with a range of about 86 millimeters. Hahn's attempt to separate this short-lived element, which he called thorium C, failed. This is hardly surprising, since it was later found to have a half-life of 3×10^{-7} seconds!

Hahn also worked on the sample of actinium that Ramsay had given him, and he had the thrill of making a discovery that had just been missed by Rutherford's first foreign student, the Polish chemist Tadeusz Godlewski. This was a new substance that fell between actinium and actinium X. Hahn named it radioactinium and showed that it decayed into actinium X in 19.5 days. Once again Rutherford expressed doubt at the proliferation of radioactive substances, but he was soon convinced that Hahn was right.

Hahn then made a careful comparison of the range of the alpha emissions of the actinium and thorium series, with the following results:

Actinum series	Range in millimeters	Thorium series	Range in millimeters
radioactinium	48	radiothorium	39
actinium B	55	thorium B	50
actinium emanation	58	thorium emanation	55
actinium X	65.5	thorium X	57
		thorium C	86

This was a first step on the road that would lead to the discovery of isotopes by Soddy some six years later.

Hahn was to become a close friend of Rutherford, whose extraordinary energy and infectious enthusiasm made life in his laboratory a constant challenge. Like many of Rutherford's students, Hahn often recalled Rutherford's laughter and the boyish delight he took in repeating the *bons mots* of fellow scientists. While Hahn was in Montreal, H. Marckwald, a German chemist who had independently discovered polonium and christened it radio-tellurium, published an article in which he acknowledged the priority of Marie Curie's discovery and withdrew the name he had proposed. He ended his paper with a quotation from *Romeo and Juliet*, "What's in a name? That which we call a rose, by any other name would smell as sweet."[4] This literally enraptured Rutherford, who went around the Macdonald Physics Building making the walls echo with these lines.

Shortly before Hahn's arrival in Montreal, John McNaughton, who taught classics at Queen's University and tried for a year to teach the same subject at McGill before fleeing back horrified to the quiet haven of Kingston, penned an amusing, if somewhat satirical, description of Rutherford: "The lecturer seemed himself a large piece of the expensive and marvellous sub-stance he was describing. Radio-active is the one sufficient term to characterize the total impression made upon us by his personality. Emanations of light and energy, swift and penetrating, cathode-rays strong enough to pierce a brick-wall, or the head of a Professor of Literature, appeared to sparkle and coruscate from him all over in sheaves."[5]

Rutherford was always larger than life, and he never departed from his native informality both in his manner of address and in his way of dressing. One day a photographer came to the institute to take a picture of Rutherford and his apparatus for *Nature*. When the plate was developed, the photographer was not satisfied with it: Rutherford was not even wearing a long-sleeved shirt with cuff-links. For the second picture, Hahn obligingly lent him the detachable cuffs that he, like many others, wore in those days. But the photographer was still dissatisfied. The cuffs did not show enough! In a third picture they were prominently displayed and Hahn recalls that he had the proud satisfaction of seeing his cuffs immortalized in *Nature*.

Hahn was also amused by the occasional visits of Sir William Macdonald, the wholesale tobacconist, who had so lavishly endowed the physics laboratory. Although he had made all his money out of tobacco, he considered smoking a filthy habit and would allow no one to smoke in his presence. Rutherford was a chainsmoker, seldom to be found without a cigarette or a pipe in his mouth, and when Macdonald called to say that he was coming round, Rutherford would rush into the laboratory with stentorian cries of "Open the windows! Put away your pipes! Hide your tobacco, Macdonald is coming!"[6]

Hahn spent most of his days as well as his evenings in the laboratory, but he was a frequent visitor at Rutherford's house. There they invariably talked shop, somewhat to the annoyance of Mrs. Rutherford, who would have been gratified if her husband and his guests occasionally listened to her performing on the piano. Hahn was also invited to the home of Professor

Cox, the director of the Macdonald Physics Building. Here science took a more entertaining turn. A gas-burning candelabrum hung from the ceiling of the living room, and when a guest knocked on the door the stopcock was turned, the guest was asked to scuff across the carpeted floor to accumulate a static charge, and then to reignite the candelabrum by means of the electric spark that jumped from his extended finger.

Life in Montreal was inexpensive. A room with breakfast and dinner could be had for 5 dollars a week, and lunch at the university cost 25 cents. But Hahn and his friend Max Levin found that meals at the boarding house hardly provided the energy needed by a research student, so they made a point of dining out at the Windsor Hotel, then the best in town, every Sunday. This cost them a dollar apiece. After a while Hahn and Levin decided to economize on the food, but they would not forsake their Sunday evening beer. When they ordered only beer, they were told that alcoholic beverages could not be served without food. "How about ordering a ham sandwich?" the waiter suggested. They complied and the waiter picked up a sandwich, slightly yellow with age, from a neighboring table and brought it to them before fetching their beer. The two German scientists watched with amusement as "their" sandwich subsequently moved from table to table as new customers arrived.

Hahn left Montreal in the summer of 1906 to take up an appointment at the Chemical Institute at the University of Berlin. But since he was not working in a conventional field of chemistry, he could not become an official assistant. Indeed, there was no room for his equipment in the laboratory, and the director of the institute, Emil Fischer, could only think of the carpentry shop in the basement as an alternative. The planing bench was removed from the shop, and Hahn installed his three electroscopes — one each for alpha, beta, and gamma rays — on an ordinary writing table. In Montreal these insturments had been made of old coffee and tobacco tins. In Berlin Hahn now had neat brass boxes and amber to use as insulation. The aluminium-foil strips were still charged, however, with sticks of hard rubber that had been vigorously rubbed on one's coat sleeves. Hahn still had his radiothorium and actinium salts, and he was able to purchase (always from Giesel in Braunschweig) 2 milligrams of radium for 200 marks. Uranium, in the form of uranyl nitrate, was then considered worthless by mine owners and could be had for a song.

Hahn returned to Germany not only with a greater knowledge of radioactivity but also with much fondness for the ease and informality of relations between younger and older research scientists at McGill. The pomposity of the world of academe had been pricked forever. This did not, however, endear him to some of his more senior colleagues in his homeland. In the spring of 1907 Hahn lectured on radioactivity at a meeting in Hamburg. The well-known physical chemist Gustav Tammann challenged him from the floor and flatly denied the existence of radium as an element. Hahn retorted with complete frankness, and perhaps with more curtness than was the custom. After the incident, his friend Max Levin urged him to be more circumspect. He had just overheard a professor describing him as "one of those anglicized Berliners."[7] In Rutherford's laboratory plain words were encouraged, and the young were not expected to keep their peace. Hahn, however, was able to see the humor of the situation. One day Fischer asked him to explain some properties of mesothorium to a professor of medicine who wished to use radiation in some experiments on bacteria. The elderly physician received him cordially and began speaking of "semithorium." When Hahn gently replied that it was called "mesothorium," the doctor put his hand on his shoulder and said: "Believe me young man, the name of the substance is semithorium."[8] Hahn did not think it worthwhile to inform him that he had actually discovered and named the substance.

Chemists were frankly puzzled by Hahn's research. Although Rutherford and Soddy had published their theory of radioactive decay several years earlier, the new conception of the atom had not seeped very deeply into the working consciousness of chemists. They found it hard to believe that the minute quantities of radioactive material were really elements, especially since their chemical characteristics could not be determined. They marvelled at the assertion that in pitchblende there is only about one part of radium to three million parts of uranium, and yet the activity ascribed to radium was several times that of uranium. The short life of some of these substances also added to the difficulty they experienced in recognizing them as honest-to-goodness "chemical" elements. These transitory and invisible substances seemed quite untrustworthy. Even Emil Fischer, who hired and helped Hahn, was surprised when Hahn read a paper on radioactive elements and stated that they were present in such small amounts that they could not be weighed but only detected by their activity. Fischer argued that even the smallest amounts of substance give off an odor, and he described his own experiments with cacodyl compounds that can be detected by smell. Hahn was able to convince Fischer and his colleagues that his nonweighable elements were genuine, but he never aroused much enthusiasm about them.

The physicists were more sympathetic, and it was at a colloquium at the Institute of Physics that he met, on November 27, 1907, the person who was to become his closest collaborator. This was Lise Meitner, one of the first women to obtain a doctorate in physics at the University of Vienna. Meitner had worked on radium in Ludwig Boltzmann's Institute for Theoretical Physics, and she had already published two papers on radioactivity before coming to Berlin to attend the lectures of Max Planck. Hahn and Meitner soon found that they had much in common, but Emil Fischer was reluctant to create a precedent by allowing a woman to enter his institute. After Planck personally intervened, he agreed to let her work in the carpentry shop but he forbade her to ascend the stairs to the laboratories on the upper floors. Reminiscing in his old age, Hahn described Lise Meitner as a shy and reserved person with whom he had no contact outside the institute or at professional meetings. He must have forgotten that, along with many other scientists, they met at Planck's house for evenings of music. Lise Meitner herself remembered the carpentry shop as a cheerful place to work: "When our work went well we sang together in two-part harmony, mostly songs by Brahms. I was only able to hum, but Hahn had an outstanding voice. Our personal and professional relations with our young colleagues at the neighboring Physical Institute were excellent. They often visited us and would occasionally climb in through the window of the carpentry shop instead of taking the usual way. In short, we were young, happy, and carefree — perhaps politically too carefree." [9]

With young men climbing in through the window, Lise Meitner cannot be said to have lived the life of a recluse, but she was for many years considered an intruder in a man's world. The readers of her articles in learned journals often assumed she was a man. One day, an editor of the well-known Brockhaus Encyclopedia wrote to the editor of a professional journal in which her articles appeared to request the address of "Mr." Meitner whom he wished to write for the encyclopedia. When informed that she was a woman, he immediately replied that his publishing house would not think of hiring a woman author! There was no female professor of any rank in Berlin. In 1912, Planck took the daring step of making her an "assistant" at the Institute of Theoretical Physics at the University of Berlin, but it was only after World War I that women were officially admitted to academic careers. Lise Meitner's inaugural lecture as a *Privatdozentin*, on "Problems of Cosmic Physics," was duly reported in the local press but with a subtle change in the title which, as "Problems of Cosmetic Physics," acquired a more feminine stamp.

I have already mentioned that upon his return to Berlin Hahn pursued the line of investigation that had given rise to his disagreement with Boltwood over the half-life of radiothorium, and that this had led him to uncover a main portion of the thorium disintegration series. As he had surmised, meso-thorium 1 (radium-228) and mesothorium 2 (actinium-228) were intermediate products between thorium (isotope 232) and radiothorium (thorium-228). The discovery of mesothorium furnished Hahn with the reason why he had discovered radiothorium in the sample of barium salts which Ramsay had given him. He had assumed that the sample contained only radium, but it had also contained mesothorium that decayed into radiothorium. Every time he made a fractional crystallization of the barium salts in order to enrich them with radium, the small amount of radiothorium that had formed in the meantime was removed. With hindsight, Hahn saw that the enrichment of radiothorium in the radium-barium solutions would have proven sisyphean since the quantity that was removed was being replenished at a constant rate.

Hahn and Lise Meitner tried to separate mesothorium from the radium, but to no avail. They suspected a strong chemical similarity such as had been observed with some of the rare-earth elements that can only be separated after a long series of careful fractional crystallizations. It did not occur to Hahn at the time that they might have the same chemical characteristics and hence belong in the same place in the periodic table.

Hahn and Meitner decided to apply the results of Rutherford's research on alpha rays to the study of beta radiation. Since facilities in the carpentry shop were inadequate, they joined forces with Otto von Baeyer of the Insti-tute of Physics. Their apparatus was very similar to that which Rutherford had constructed at McGill. The radioactive material was collected on a thin wire 0.2 millimeters in diameter and 25 millimeters long. This wire was placed in a narrow groove at the bottom of the apparatus. Above this a slot allowed a narrow beam of radiation to reach a photographic plate. The bending of the beta rays could be determined from the photographs with the aid of a low-magnification telescope. Hahn and Meitner had initially assumed that each beta line represented a separate substance, but they soon realized that what they had measured was not the absorption but the dispersion. The greater the distance of the preparation from the bottom of the electroscope, the greater the dispersion. By increasing the distance, they had dispersed the weakest beta rays so much that they failed to register. This work laid the foundation, however, of Meitner's subsequent distinction between secondary beta rays, with their sharp magnetic spectra, and primary beta rays from atomic nuclei which produced the weak lines.

During 1908 and 1909 Hahn detected the radioactive recoil of alpha particles that Rutherford had conjectured as an explanation of apparent inconsistencies in the behaviour of radium B at normal temperatures. Harriet

Brooks, one of Rutherford's first students at McGill, had investigated the problem, but she failed to offer a complete explanation since she believed that the recoiling particles could not be concentrated on the negative electrode but were regularly distributed in an electrical field. Hahn was able to show that they carried a positive charge and to offer a satisfactory interpretation of their production.

At the large centennial celebration of the University of Berlin in 1910 Kaiser Wilhelm II unveiled a plan to create major research institutes. It was agreed that the Kaiser Wilhelm Institute for Chemistry would be the first to be erected. This was to prove a decisive event in the lives of Otto Hahn and Lise Meitner. Hahn was immediately invited to head his own small department and Meitner joined him shortly thereafter. The building was officially opened on October 12, 1912. Since the Kaiser was to attend the ceremony, Hahn was asked to mount a radioactive display for him. Hahn decided to repeat the successful exhibition that he had put on for ladies at the annual social gathering of the Royal Society in London when, as a student of Ramsay's, he had showed the effects of the emanation of his radiothorium on a fluorescent screen placed in a dark room. This had thrilled his audience, for upon entering the room they could see nothing until their eyes became adjusted to the darkness and they began seeing luminous shapes moving on the screen. It was agreed that such a display might interest the Kaiser, but an unexpected difficulty arose on the evening of the official opening. An aide-de-camp to the Kaiser came to inspect the building, and when Hahn attempted to lead him into the darkened room to show him the radioactive preparations, he burst out indignantly: "Out of the question! We cannot send His Majesty into a completely darkened room."[10] A compromise was found and a small red lamp was installed. When the Kaiser arrived the next day, he did not show the slightest hesitation to enter the darkened room, and everything went as planned.

Research was interrupted by the outbreak of World War I. Hahn was immediately drafted and ordered to report to Fritz Haber, the head of the Kaiser Wilhelm Institute for Physical Chemistry, who was also the inventor of the high-pressure ammonia-synthesis process that was to have such a baneful influence during the war. Hahn later recounted that Haber told him that "the fronts which had hardened in the West were to be surmounted only by the use of new weapons, particularly noxious and poisonous gases, above all chlorine gas, that would have to be released on the enemy from the forward positions. To my objection that this kind of warfare violated the Hague Convention, he replied that the French had already started it, although weakly in the form of gas-filled artillery shells. Moreover, he said, numerous lives would be saved if the war could be ended faster in this way."[11]

Hahn was sent to every front. In Poland he once led an attack using a mixture of chlorine and phosgene. During the ensuing advance he met several gassed Russians.

I felt profoundly ashamed. I was very much upset. First we attacked the Russian soldiers with our gases, and then when we saw the poor fellows lying there, dying slowly, we tried to make breathing easier for them by using our own life-saving devices on them. It made us realize the utter senselessness of war. First you do your utmost to finish off the stranger over there in the enemy trench, and then when you're face to face with him you can't bear the sight of what you've done and you try to help him. But we couldn't save those poor fellows.[12]

Fritz Haber was a victim of circumstances, and Hahn never begrudged him respect and admiration even after he had been designated a "war criminal" abroad and persecuted as a Jew in Nazi Germany. Haber emigrated, a broken man, and died in Basel in 1934. A year later, in the face of considerable opposition, Max Planck, the president of the Kaiser Wilhelm II Society, decided to hold a commemorative ceremony. Participation by university professors was formally prohibited by the Minister of Education. Hahn had already resigned his post at the University of Berlin, but he took a grave risk when he accepted to give the address at the memorial service.

After the war, Hahn and Meitner resumed their work at the Kaiser Wilhelm II Institute. In 1918 they were able to announce the discovery of protactinium, the mother substance of actinium, number 91 in the series of elements. A few years later, they discovered uranium Z, the first instance of a nuclear isomer. Hahn was fond of this as an example of how one sometimes finds something quite different from what one is looking for. A lucid account of this discovery is offered by Ernst H. Berninger in his contribution to this volume.

In 1919 Rutherford astounded the scientific world by announcing a change in the nucleus of an atom. He bombarded nitrogen atoms with alpha particles — that is, helium nuclei — by exposing the atoms to a highly active preparation of radium. The nitrogen nucleus is considerably heavier than the alpha particle and so should not be driven forward much as the result of being struck. Nevertheless, Rutherford found that when alpha particles passed through nitrogen, many long-ranged particles were produced, as they were when the particles passed through hydrogen. With oxygen, whose mass does not differ much from that of nitrogen, no long-range particles occur.

He further showed, by the use of a magnetic field, that the long-range particles from nitrogen behave like hydrogen nuclei. After several checks he was
driven to the conclusion that the nitrogen nucleus had been disintegrated
by the impact of an alpha particle and a hydrogen nucleus had been knocked
out of it. It occurred to Rutherford that what actually took place in the
collision might be recorded by the cloud chamber method, which would
show the path of the alpha particles before impact and the paths of the
resultant particles after impact. One of his co-workers, Patrick Blackett,
built a cloud chamber apparatus that automatically took a photograph every
15 seconds. In a few months in 1924 Blackett took some 23,000 photographs
of alpha tracks in nitrogen. There were, on average, eighteen tracks on each
photographic plate, so that he had in all some 400,000 tracks. Among these
he found eight branched tracks of a new character, one branch being a very
long thin track, shown to be due to a proton (the word Rutherford introduced for the nucleus of hydrogen), and the other a short, thick branch,
resulting from an oxygen isotope formed after impact.

The first artificial disintegration of a nucleus opened a new era in physics.
In 1932 James Chadwick discovered the neutron, which had been predicted
by Rutherford in his Bakerian Lecture of 1920. In 1933 isotopes that were
radioactive were artificially made by the Joliot-Curies. Immediately afterward
Enrico Fermi and his collaborators in Rome bombarded stable elements with
neutrons and produced, out of sixty-three elements investigated, thirty-seven
new elements showing radioactive qualities. By this time the Hahn-Meitner
team in Berlin had been increased by a third member, Fritz Strassmann,
who was to play a key role in subsequent developments. Lise Meitner, as an
Austrian citizen, had been able to avoid the anti-Semitic laws, but after the
annexation of her country to the Third Reich on March 13, 1938, she was
seriously threatened, and her friends arranged for her to leave for Sweden.
Hahn was deeply distressed. He had lost his closest collaborator, he was
considered "unreliable" by his own government, and fundamental research
was increasingly under attack in Germany. He sought refuge in his laboratory
and with Strassmann performed his experiments with a new sense of urgency
— perhaps these would be the last that he would be allowed to do. They
irradiated uranium with neutrons to find which new elements were formed.
The story and the significance of this quest is brilliantly analyzed by Fritz
Krafft and Spencer R. Weart in their essays on the discovery of nuclear
fission.

In brief, the answer that Hahn and Strassmann found was that radium
comes from uranium. Physicists were skeptical, since only transformations

of an atom into a neighboring atom had hitherto been obtained. How could one pass from uranium, number 92 in the periodic table, to radium, number 88, without making intermediate stops? To work with such a small amount of radium, a "carrier substance" was required and barium was chosen, because it is chemically related to radium, and hence radium remains closely bound to the carrier in precipitation. The results were perplexing. The finest analyses always yielded the carrier substance barium! But how could barium (atomic number 56) arise from the irradiation of uranium (atomic number 92)? This was only possible if the atom had actually been split. Hahn was anxious to hear what Meitner thought of the suggestion. She mulled over the possibilities with her nephew Otto Robert Frisch. If Hahn were right, then barium is a piece of the uranium nucleus and the other piece must have the atomic number 36 (92 − 56 = 36); that is, it must be krypton.

Within a short period of time, laboratories all over the world confirmed the Hahn−Strassmann experiment. It soon became clear that a large amount of energy is freed at each splitting of the uranium nucleus. The process initiated by one neutron is associated with the simultaneous production of two to three neutrons. According to this result, it should be possible to set up a chain reaction and to create an explosive device of unheard violence or, if the reaction could be tamed, a source of fantastic energy. While these new ideas were being discussed by physicists, German troops invaded Prague. The fragile peace was nearing an end, and for the next six years the world was plunged into armed conflict. The climate of opinion at the outbreak of the war is discussed by Neil Cameron in his essay on the politics of British science in the Munich era, and Lawrence Badash provides an illuminating account of the genesis of Hahn's social and moral stance.

During the war, Hahn continued to work in Berlin and did not engage in Heisenberg's uranium project. After the defeat of Nazi Germany, he was taken prisoner − along with nine other German scientists − and interned at Farm Hall near Cambridge. It is here that they learned on August 6, 1945, of the explosion of an atomic bomb over Japan. None of the ten scientists could sleep that night. Max von Laue, a close friend of Hahn's, was deeply worried about Hahn's obvious distress at such a lethal use of his discovery of atomic fission. The internment at Farm Hall became senseless after Hiroshima, but the German scientists were detained until the end of the year. While still in captivity, Hahn received one item of good news: he had been awarded the Nobel Prize. Upon his return to Germany he was instrumental in rebuilding the Kaiser Wilhelm Society of which he became president when it was officially refounded as the Max Planck Society in 1948. He became increasingly worried about the danger of a nuclear holocaust, and on February 13, 1955, he made an important radio broadcast on "Cobalt 60: Danger or Blessing for Mankind." The international reaction was encouraging, and he decided to organize the joint statement by Nobel Prize winners that came to be known as the Mainau Declaration. It reads in part:

> We are natural scientists from various countries, various races, various beliefs, various political convictions. We are bound externally only by the Nobel Prize that we have all received.
>
> We have happily placed our lives in the service of science. It is, we believe, one route to happier life for mankind. We see with shock that this very science also provides mankind with the means to destroy itself.
>
> An all-out use in war of the weapons possible today can contaminate the earth with such radioactivity that entire nations would be annihilated. Neutral countries would share in this death just as the belligerents.
>
> If a war were to break out between the great powers, who could guarantee that it would not develop into such a death struggle? Thus a nation that opens the door to total war brings about its own destruction and endangers the entire world.
>
> We do not deny that perhaps peace today is preserved precisely because of fear of these deadly weapons. Nevertheless, we regard it to be self-delusion if governments believe they can avoid a war for a long period of time only by fear of these weapons. Fear and tension have so often produced war. Likewise, it appears to us a self-delusion to believe that small conflicts will continue to be decided by traditional weapons. In extreme danger no nation will deny any weapon that scientific technology can produce. All nations must decide voluntarily to refrain from violence as the last means of politics. If they are not prepared to do so, they will cease to exist.[13]

Hahn lived to be eighty-nine and died in 1968, but these words from the Mainau Declaration may be taken as his spiritual testment, a message that is as vitally important today as it was when it first appeared.

The opening essays in this volume, the first by Roger H. Stuewer on the nuclear electron hypothesis and the second by Robert Kargon on the research program of Robert Millikan, present an authoritative picture of some of the more fascinating aspects of physics before the discovery of nuclear fission. The concluding paper by B. W. Sargent offers a useful summary of work done in Canada in the 1930s.

McGill University

NOTES

[1] Ernst H. Berninger, *Otto Hahn in Selbstzeugnissen und Bilddokumenten* (Reinbek bei Hamburg: Rowohlt, 1974), p. 28.

[2] Otto Hahn, *My Life*, trans. Ernest Kaiser and Eithne Wilkins (New York: Herder and Herder, 1970), p. 78.
[3] E. N. da C. Andrade, *Rutherford and the Nature of the Atom* (Garden City: Doubleday, 1964), p. 56.
[4] Otto Hahn, *A Scientific Autobiography*, trans. and ed. Willy Ley (New York: Scribner's, 1966), p. 33.
[5] *McGill Univ. Mag.* **3** (Apr. 1904), 17–18.
[6] Andrade, *Rutherford*, p. 57.
[7] Hahn, *Scientific Autobiography*, p. 68.
[8] *Ibid.*, p. 70.
[9] Armin Hermann, *The New Physics* (Bonn/Bad Godesberg: Inter Nations, 1979), p. 41. I have slightly altered the rendering from the German.
[10] *Ibid.*, p. 46.
[11] *Ibid.*
[12] Hahn, *My Life*, p. 132.
[13] Hermann, *New Physics*, pp. 131–133.

ROGER H. STUEWER

THE NUCLEAR ELECTRON HYPOTHESIS

INTRODUCTION

James Chadwick's discovery of the neutron in 1932 is justifiably viewed as a watershed in the history of nuclear physics.[1] It opened the way to all modern neutron-proton theories of nuclear structure and (so the story goes) simultaneously expelled the electron from the nucleus where it had become, through its increasingly contrary behavior, an embarrassing guest — part of a problem not to be thought about, like the new taxes, as Peter Debye put it in 1930.[2] Chadwick's discovery reintroduced order and simplicity into nuclear theory.

History, however, is seldom so neat, and the life cycle of the nuclear electron hypothesis is far more interesting and complex than the above suggests. Even Edward M. Purcell's pioneering account[3] leaves a number of questions unanswered. For example, Purcell implicitly assumes that the electron was indeed expelled from the nucleus by Chadwick's discovery; yet this conflicts with Chadwick's own preference for retaining it, as we shall see, and with Heisenberg's ambivalence on the question in his seminal papers of 1932,[4] as Joan Bromberg and others have pointed out.[5]

Was, then, the electron truly an embarrassing guest inside the nuclear mansion? How did its masters — contemporary physicists — cope with its errant behavior? Through what doors, indeed, did it actually enter and actually leave? These are some of the principal questions I shall address, though limitations of space will prevent me from going into detail at many points. My main goal is to establish a framework for future research, by providing some insight into the richness and complexity of nuclear physics in the period 1911–1934.

THE ORIGIN AND FRUITFULNESS OF THE NUCLEAR ELECTRON HYPOTHESIS, 1911–1919

Ernest Rutherford's 1911 scattering theory,[6] based upon Hans Geiger and Ernest Marsden's 1909 experiments,[7] suggested that the atom, of radius about 10^{-8} centimeters, consists of "a strong positive or negative central

19

William R. Shea (ed.), Otto Hahn and the Rise of Nuclear Physics, 19–67.
Copyright © 1983 by D. Reidel Publishing Company.

charge concentrated within a sphere of less than 3×10^{-12} cm radius," as Geiger and Marsden put it in 1913.[8] The ambiguity in sign arose because Rutherford's scattering formula varied as the square of the central charge or nucleus, as it was soon called. It disappeared immediately if, as Rutherford believed, the surrounding charge was negative or consisted of negatively charged electrons.

The question of the internal constitution of the nucleus first surfaced at an unexpected location. Antonius van den Broek (1870–1926), a lawyer and amateur scientist living in Gorssel, Holland, had become interested in the problem of the periodicity of the elements, and at the end of 1912 concluded that the total number of electrons in an atom is equal to its atomic number Z or ordinal position in Mendeleev's table.[9] Niels Bohr cited van den Broek's hypothesis approvingly in Part II of his famous trilogy, transmitted to Rutherford for publication on June 10, 1913.[10] The preceding April, however, Geiger and Marsden had reported further scattering experiments[11] which had attracted van den Broek's attention. Subsequently, in November 1913 he observed that Geiger and Marsden's new data agreed much better with Rutherford's scattering formula if the nuclear charge of the scatterer were set equal not to half its atomic weight A, as Geiger and Marsden had assumed, but to its atomic number Z.[12] Furthermore, van den Broek concluded, if the nucleus consists mostly of doubly charged alpha particles, "then the nucleus too must contain electrons to compensate this extra charge."[13] Van den Broek's conclusion constitutes the first explicit statement of the nuclear electron hypothesis in the literature.

We see that part of his argument was based on the assumption, generally accepted by 1913, that alpha particles emitted in radioactive alpha decay are actually present in the nucleus as discrete particles prior to emission. The parallel argument for beta rays, which might have led directly to the nuclear electron hypothesis, was not at all obvious. The principal difficulty was that in 1911–1912 Otto von Baeyer of the Physikalisches Institut of the University of Berlin, collaborating with Otto Hahn and Lise Meitner in the Chemisches Institut, had found that the beta spectrum of thorium active deposit and other beta emitters consists of groups of beta rays of definite velocities.[14] These results were confirmed by J. Danysz in Marie Curie's laboratory in Paris,[15] and by mid-1912 H. G. J. Moseley, in Rutherford's laboratory in Manchester, had found evidence indicating that only one beta ray is emitted per atom of radium per disintegration.[16] Taking everything together, Rutherford in August 1912 suggested that a distinction could be drawn between "the instability of the central nucleus and the instability of the electronic distribution. The former type of instability leads to the expulsion of an α-particle, the latter to the appearance of β and γ-rays. . . ."[17]

Kasimir Fajans and O. W. Richardson supported Rutherford's distinction, but van den Broek and Bohr, in 1913, objected.[18] By that time, however, Rutherford himself had revised his ideas, because he and H. Robinson had found that the beta and gamma rays from radium emanation produced far too much heat energy to have had their origin in the electronic distribution.[19] At the end of 1913 Frederick Soddy, Rutherford's former McGill collaborator now in Glasgow, also pointed out that the uranium nucleus had to contain at least six electrons, since it expels six beta rays in its decay chain. This proved, he wrote, that the "central charge of the atom in Rutherford's theory cannot be a pure positive charge, but must contain electrons, as van der Broek [sic] concludes."[20]

Soddy's challenge drew an immediate and haughty denial from Rutherford that he was excluding electrons from the nucleus. He noted, in addition to his and Robinson's experiments, the fact that beta emission, like alpha emission, was "independent of physical and chemical conditions."[21] He also spoke approvingly of van den Broek's hypothesis, Bohr's use of it, and Moseley's recent X-ray experiments as confirming evidence. It is amusing to note that van den Broek, the amateur scientist, was evidently horrified to think that he might unwittingly have driven a wedge between the famous former collaborators, for he immediately stated in *Nature*: "My letter was . . . not supposed by me to give any rectification of the theory of the positive nucleus as proposed by Prof. Rutherford. Nor did I suppose the idea that the nucleus might contain electrons to be new."[22]

Far from disputing van den Broek's priority, however, Rutherford adduced still further evidence for the Dutchman's nuclear electron hypothesis. It accounted for the great stability of the alpha particle, as he explained in February 1914,[23] owing to the reduction in mass associated with the overlapping electric fields of its tightly packed positive and negative electrons, consistent with Lorentz's views. It accounted for the radioactive displacement law, associated here with beta decay, as enunciated in 1913 by Fajans and Soddy.[24] It was consistent with the possibility that "a number of atoms may exist with the same nucleus charge but of different masses,"[25] that is, with the existence of isotopes, as they soon were called. Finally, Rutherford reexamined his views following Chadwick's discovery, made while working in Geiger's Berlin laboratory in 1914 shortly before he was interned for the duration of the war, that the beta-ray spectrum of radium B + C consists of a *continuous* spectrum, upon which is superimposed the *line* spectrum.[26]

Rutherford offered a statistical interpretation of the former and a quantum interpretation of the latter. He assumed fundamentally that beta rays are emitted from the nucleus with discrete velocities, and that any gamma rays produced originate in the vibration of the inner electrons and are not themselves nuclear radiations.[27] Two decades would pass before the origin of alpha, beta, and gamma rays would be fully understood.

Nuclear theory, in fact, was in such an embryonic state prior to World War I that Rutherford in 1913 proposed leaving it to the next generation of physicists.[28] His pessimism was tragically reinforced after the guns of August sounded: Moseley was killed at Gallipoli, Chadwick was interned in Berlin, Hahn was drafted into the *Landwehr*, Meitner volunteered to serve as X-ray technician in the Austrian army, Geiger became an artillery officer at the front, Marie Curie began her legendary war service, and Rutherford himself became firmly drawn into anti-submarine work.[29]

Despite these enormous disruptions in experimental research, speculation flourished. Between 1915 and 1919 specific nuclear models were proposed by H. S. Allen,[30] W. D. Harkins,[31] A. E. Haas,[32] J. W. Nicholson,[33] J. J. Thomson,[34] A. W. Steward,[35] Emil Kohlweiler,[36] Hans Th. Wolff,[37] Walther Kossel,[38] Torahiko Terada,[39] E. Gehrcke,[40] and Ingo W. D. Hackh[41] − a truly international cast of physicists and physical chemists. Although I cannot discuss these models in detail, I wish to emphasize that all assumed the existence of electrons in the nucleus, proving beyond doubt the reasonableness and pervasiveness of that hypothesis. I might also note that it appears that physical chemists were particularly attracted to nuclear speculation during this period and that their imaginations were particularly unbridled. Two examples will have to suffice. First, the Chicago physical chemist W. D. Harkins in 1915 began a program of numerological speculation on isotopic structures seldom equaled in the history of science. A paper written in 1919, for example, contained seventeen "postulates" for specifying the structure of the twenty-seven lightest nuclei, as well as the model of the alpha particle shown in Figure 1, complete with nuclear electrons in "the form of rings, or disks, or spheres flattened into ellipsoids."[42] I agree entirely with E. M. Purcell's characterization of Harkins's work as containing "no more than a description of the empirical regularities of the isotope chart."[43]

As a second example, the Stuttgart physical chemist Emil Kohlweiler in 1918 produced quite likely the most spectacular nuclear model of the period, one comparable to a Gothic cathedral, illustrated in Figure 2 for the case of atomic number 44, atomic weight 118.[44] We see that the nucleus (*Atomkern*) consists of positive and negative *groups* − note the nuclear electrons − each

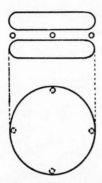

Fig. 1. Harkins's model of the alpha particle. The four "positive electrons" are placed symmetrically between the two ellipsoidal negative electrons. ("The Nuclei of Atoms and the New Periodic System," *Phys. Rev.*, 15 (1920), 82. Reproduced by permission of the Department of Chemistry, University of Chicago.)

of eight unit charges. I will not describe the contents of the middle zone or outer zone, nor explain how the model was supposed to account for radioactive decay, chemical bonding, and the periodicity of the elements.

SOME CONSEQUENCES OF RUTHERFORD'S DISCOVERY
OF ARTIFICIAL DISINTEGRATION, 1919–1929

Experiment has always imposed constraints on the imagination, and Rutherford's discovery of artificial disintegration in early 1919,[45] just before he left Manchester for Cambridge, exerted a decisive influence on all subsequent theorizing on nuclear structure. This discovery, the culmination of years of sporadic wartime researches carried out with the assistance of William Kay, his Manchester laboratory steward, proved that radium C alpha particles incident on nitrogen give rise to long-range secondary particles (maximum range about 28 cm in air), which were "probably atoms of hydrogen." "If this be the case," Rutherford wrote, "we must conclude . . . that the hydrogen atom which is liberated formed a constituent part of the nitrogen nucleus."[46] Repeating and extending his experiments at the Cavendish with Chadwick, Rutherford was in a position to summarize and interpret his work in his second Bakerian Lecture, "Nuclear Constitution of Atoms," which he delivered on June 3, 1920, before the Royal Society of London.[47] Everyone agreed with Bohr that Rutherford's discovery opened up "a new epoch in natural philosophy."[48] Arnold Sommerfeld, in particular, took pains to

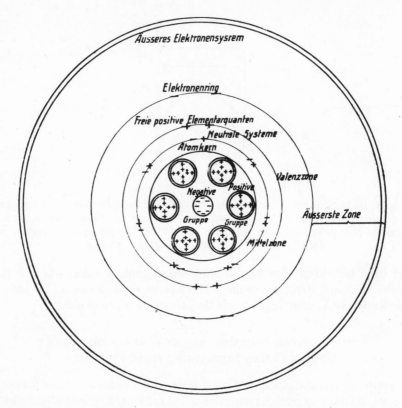

Fig. 2. Kohlweiler's atomic model for atomic number 44, atomic weight 118. Judging from modern data, this seems to be a wholly fictitious example. ("Der Atombau auf Grund des Atomzerfalls und seine Beziehung zur chemischen Bindung . . . ," *Z. phys. Chem.*, **93** (1918), 11. Reproduced by permission of the Akademische Verlagsgesellschaft, Wiesbaden.)

add a special section on it in the first edition (1919) of his famous *Atombau und Spektrallinien*, in which he proved from Einstein's mass-energy relationship and mass-defect considerations that the alpha particle was a much more stable structure than the nitrogen nucleus and hence ought to survive the collision process intact.[49]

Rutherford's conclusion that hydrogen nuclei (soon called protons) had to join electrons and alpha particles as fundamental nuclear constituents led him to suggest a specific model for the nitrogen nucleus in his first publication of June 1919. He suggested that two hydrogen nuclei were "outriders of the main system of mass 12" and were at a distance of "about twice the diameter of the electron (7×10^{-13} cm) from the centre of the main atom."[50] The incident alpha particles, then, as he explained in detail in 1921,[51] would strike these outriders at various points in their orbits, as shown in Figure 3, and expel them with their observed velocity distribution. That the alpha particle

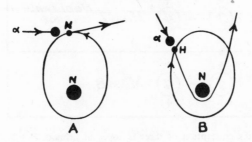

Fig. 3. Rutherford's conception of the disintegration of the nitrogen nucleus by alpha particles, ejecting hydrogen nuclei. ("The Artificial Disintegration of Light Elements," *Phil. Mag.*, 42 (1921), 822; reprinted in *Collected Papers of Rutherford*, Vol. III, p. 60. Reproduced by permission of Taylor & Francis, Ltd.)

was actually *captured* by the nitrogen nucleus, and hence that the residual nucleus was actually an isotope of *oxygen*, not one of carbon as Figure 3 suggests, became clear only after P. M. S. Blackett obtained cloud chamber photographs of the disintegration process at the end of 1924.[52]

On such a model, Rutherford noted in his Bakerian Lecture, the nuclear electrons were necessarily subjected to "intense forces" and were "much deformed" when incorporated into stable nuclear subsystems.[53] One such subsystem was the alpha particle, composed of four protons and two electrons. Another was a new particle which Rutherford believed he had detected as a disintegration product of both oxygen and nitrogen – a combination of three protons and one electron, an X_3^{++} particle. While Rutherford and Chadwick proved conclusively in 1924 that this new particle was actually an ordinary alpha particle of 9.3 centimeter range emitted by the radium C source,[54] in 1920 Rutherford took it to be a major new discovery, and it entered into his thinking in a fundamental way.

We find, in the first place, that it led him to propose definite (though he insisted "purely illustrative") nuclear structures for three isotopes of lithium, as shown in Figure 4, and for $^{12}_{6}$C, $^{14}_{7}$N, and $^{16}_{8}$O, as shown in Figure 5.[55] In these figures one can see how the nuclear electrons (−) function as

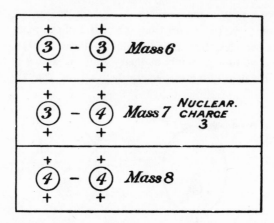

Fig. 4. Rutherford's conception of three isotopes of lithium, $Z = 3$, $A = 6$, 7, and 8. ("Nuclear Constitution of Atoms," *Proc. Roy. Soc.*, A97 (1920), 398; reprinted in *Collected Papers of Rutherford*, Vol. III, p. 35. Reproduced by permission of the Royal Society.)

particles binding together the hydrogen nuclei, alpha particles, and X_3^{++} particles. The extent to which Rutherford viewed these models as pictures of physical reality perhaps may be judged from the fact that he constructed them out of red and white balls (positive and negative particles) for a lecture at a meeting of the British Association a few months later.[56]

But Rutherford's new X_3^{++} particle entered into his thinking in a second fundamental way: it served as his point of departure for his most general speculations on the internal constitution of nuclei. If one electron can bind three hydrogen nuclei, he conjectured in his Bakerian Lecture, "it seems very likely that one electron can also bind two H nuclei and possibly also one H nucleus."[57] The latter would represent a tightly bound combination, forming "a kind of neutral doublet." And Rutherford then went on to describe the "very novel properties" of this neutral doublet which followed from its "practically zero" external field. He also noted that such a particle would be essential for building up the heavy nuclei.

This of course constitutes Rutherford's famous prediction of the existence of the neutron, and several remarks are in order regarding it. First, the influence of W. H. Bragg's theory of gamma rays and X rays as corpuscular neutral doublets,[58] which Bragg developed between 1907 and 1912, is clearly evident and was soon explicitly acknowledged.[59] Second, Rutherford's unsuccessful search for the neutron, begun immediately by two of his research

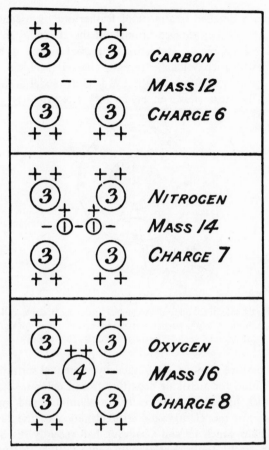

Fig. 5. Rutherford's conception of isotopes of carbon ($Z = 6$, $A = 12$), nitrogen ($Z = 7$, $A = 14$), and oxygen ($Z = 8$, $A = 16$). ("Nuclear Constitution of Atoms," p. 399; reprinted in *Collected Papers*, Vol. III, p. 37. Reproduced by permission of the Royal Society.)

students, J. L. Glasson and J. K. Roberts,[60] and continued throughout the 1920s by Chadwick,[61] exerted a profound influence on Chadwick's thought, as we shall see. Finally, if we are to agree with E. N. da C. Andrade that Rutherford's prediction merits comparison with Helmholtz's remark about Faraday, "er riecht die Wahrheit,"[62] we must also recognize that, as in so many instances in the history of physics, "die Wahrheit war auf Unwahrheit aufgebaut."

I must defer a detailed treatment of Rutherford's subsequent views on nuclear structure, as they evolved throughout the 1920s in tandem with his and Chadwick's experiments. He definitely proved himself to be a "crypto-theoretician," as Maurice Goldhaber recently characterized him,[63] developing his hydrogen "outriders" into a quantitative satellite model of the nucleus, by 1925 reaching the stage shown in Figure 6.[64] The satellites, now negative

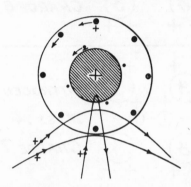

Fig. 6. Rutherford's satellite model of the nucleus showing positive and negative satellites and the trajectories of alpha particles of various energies incident upon it. ("Studies of Atomic Nuclei," *Engineering,* 119 (1925), 438.)

electrons and positive protons, established two potential surfaces of opposite sign and accounted for a striking departure from Coulomb scattering which he and Chadwick had observed recently for aluminum and magnesium. At the same time, this model suggested an explanation of the paradoxical behavior of uranium which he and Chadwick had recently discovered — that while uranium freely *emits* relatively low energy alpha particles, of range 2.7 centimeters, ones of much higher energy, of range 6.7 centimeters, cannot *penetrate* the uranium nucleus but are simply scattered away in approximately normal fashion. Rutherford concluded that this striking paradox could be understood if the negative and positive satellites — now electrons and alpha particles — were closely spaced (separation about 3 × 10^{-13} cm) to form essentially "neutral doublets." High-energy incident alphas would then be scattered normally by the nuclear core, whereas low-energy alphas would be emitted when the neutral satellites break up. Encouraged by a subsequent proof by P. Debye and W. Hardmeier that polarization forces could hold such neutral satellites in equilibrium about the core,[65] Rutherford developed his model quantitatively in 1927, showing, *inter alia,* that a large number ,of gamma-ray lines could be interpreted as arising from quantum transitions of such satellites.[66]

This small window into Rutherford's thought during the 1920s shows above all how firmly the nuclear electron hypothesis became embedded into his thinking, and how firmly he became committed to his specific satellite model of the nucleus. Indeed, he was reluctant to abandon it even after he became aware of, and understood, George Gamow's quantum theory of alpha decay in the fall of 1928, which completely resolved the uranium paradox above.[67] Thus we find the remarkable situation that in *Radiations from Radioactive Substances*, the famous treatise by Rutherford, Chadwick, and C. D. Ellis published in 1930, Gamow's theory is discussed at one point and Rutherford's satellite model at another — followed, however, by highly critical comments on the latter written by Chadwick, which were included over Rutherford's grunts.[68]

In addition to the Cavendish Laboratory, other European laboratories, overcoming the ravages of war and inflation, became active once again after the Armistice — in Vienna, Berlin, and Paris, not to mention the theoretical centers in Copenhagen, Munich, and Göttingen. Rebirth brought competition, even outright conflict. Hans Pettersson and Gerhard Kirsch, working in Stefan Meyer's Institut für Radiumforschung in Vienna, beginning in 1922 carried on a running dispute with Rutherford and Chadwick over their alpha disintegration work — a dispute which was not stalemated until Chadwick visited Vienna at the end of 1927 and found definite indications of psychological bias on the part of the Vienna observers.[69] Concurrently, Lise Meitner, working in the Kaiser-Wilhelm Institut für Chemie in Berlin-Dahlem, challenged C. D. Ellis's interpretation of the continuous beta-ray spectrum, one of the most vexing problems ever to confront physicists. Decisive new information bearing on this dispute was provided by Ellis and W. A. Wooster's 1927 calorimetric proof that beta-rays emerge from the nucleus with a continuous energy distribution [70] — a fact confirmed in more sophisticated experiments by Meitner and W. Orthmann at the end of 1929.[71] By then, as Leonard B. Loeb remarked, time had shown that both had been "in part right and in part wrong."[72]

THE PERVASIVENESS OF THE NUCLEAR ELECTRON HYPOTHESIS DURING THE 1920s

Common to all parties concerned in these important controversies was an unquestioned acceptance of the nuclear electron hypothesis and a firm reliance on their own experimental data. By contrast, there was a vast amount of speculation on nuclear structure by others during the 1920s that was often only loosely connected, if at all, to their own experimental work. For those brave enough to wade in this swamp, a bird's-eye view may be obtained by consulting Chapter 6 in A. F. Kovarik and L. W. McKeehan's 1925 National Research Council report on radioactivity,[73] whose opening section is entitled "Speculations as to Nuclear Structure." Kovarik and McKeehan cite an international cast of nine authors or coauthors who by mid-decade had proposed formulae "for nuclei regarded as pseudo-chemical compounds of protons, electrons, and relatively stable groups of these, analogous to chemical radicals."[74] Six others had proposed formulae of "the same general type." Both Harkins and Meitner had suggested that alpha particles neutralized by two electrons should be regarded as a subunit in radioactive nuclei, a suggestion extended to all elements by M. C. Neuburger but criticized by F. P. Valeras. Eight authors had proposed "extremely complicated" theories based upon classical electrodynamics and assuming definite spatial configurations for the nuclear constituents. W. Lenz, O. Stern and M. Volmer, and Y. Takahashi had proposed inverted Bohr models − protons rotating about electrons − while fifteen other authors or coauthors had advanced "more artificial" theories depending upon "special assumptions." Finally, seven authors had carried out atomic weight calculations related to the above theories. Andrade went to the heart of the matter when he observed that "the subject has offered a vast field for what the Germans call *Arithmetische Spielereien*, which serve rather to entertain the players than to advance knowledge."[75]

Fortunately, at least for the present, I need not become a participant, though I have examined many of the papers mentioned. My basic point, that the nuclear electron hypothesis pervaded physics during the 1920s, can be demonstrated in another way, by referring to standard textbooks of the period.

To begin with R. A. Millikan's book *The Electron*, we find that while in the first edition of 1917 he regards the nuclear electron hypothesis as "plausible ... yet speculative," in the second edition of 1924 he simply asserts, in italics, that *"the atomic weight [A] minus the atomic number [Z] gives at once the number of negative electrons which are contained within the nucleus of any atom."*[76] To Arnold Sommerfeld, in the 1919 first edition of his enormously influential *Atombau und Spektrallinien*, it was a simple mathematical problem: for a nucleus of given A and Z, one could construct it out of x alpha particles, y hydrogen nuclei, and z electrons, provided only that the Diophantine equations

$$A = 4x + y$$
$$Z = 2x + y - z$$

continued to be satisfied.[77] I have already noted how Sommerfeld used mass-defect considerations to prove that the alpha particle is much more stable than the nitrogen nucleus. In this connection he was much taken with Lenz's 1918 model of the alpha particle — four hydrogen nuclei revolving in an equatorial plane with one electron at each pole — a model that he presented again in the second edition of *Atombau* of 1921, the third of 1922, and the fourth of 1924. By the fifth edition of 1931 he was still firmly committed to the nuclear electron hypothesis.[78]

F. W. Aston, in the first and second editions of his classic work *Isotopes* (1922, 1924), illustrated various isotopes (same Z, different A) and "isobares" (different Z, same A), as shown in Figure 7.[79] We see here the "standard bricks" — protons and electrons — inside the nucleus (dark circle), and the "planetary electrons" outside. At another point Aston also gave a table of various light isotopes in which the nuclear electrons and protons were indicated by open circles and by dark dots, respectively, as shown in Figure 8. Aston's book, in general, is filled with insights, but I will restrict myself to quoting a single remark which was frequently repeated later, for example by Marie Curie, and which reveals Aston's skepticism regarding the presumed pre-existence of alpha particles inside radioactive nuclei: "In the writer's opinion, this is much the same as saying that a pistol contains smoke, for it is quite possible that the α particle, like the smoke of the pistol, is only formed at the moment of its ejection."[80] It was often forgotten later that Aston actually made this remark with reference to alpha not beta-decay.

The nuclear electron hypothesis also formed an integral part of the discussions on nuclear structure in N. R. Campbell's 1923 *Structure of the Atom*; of the first, second, and third editions (1923, 1924, 1927) of E. N. da C. Andrade's work of the same title; of Georg von Hevesy and Fritz Paneth's *Lehrbuch der Radioactivität* (1923; 1st English edition 1926); of Jean Perrin's *Atoms* (2nd English edition, 1923); of Hans Pettersson and Gerhard Kirsch's *Atomzertrümmerung* (1926); of Pettersson's *Künstliche Verwandlung der Elemente* (1929); and, of course, of Rutherford, Chadwick, and Ellis's *Radiations from Radioactive Substances* (1930).[81] In addition to these books, I could also cite numerous examples drawn from the periodical literature proving the pervasiveness of the nuclear electron hypothesis throughout the 1920s. I will confine myself to a single instance. In 1928 W. D. Harkins, during the course of his voluminous epistles, drew a logical conclusion: that

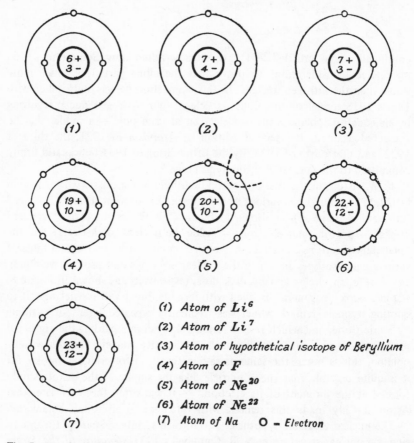

(1) Atom of Li^6
(2) Atom of Li^7
(3) Atom of hypothetical isotope of Beryllium
(4) Atom of F
(5) Atom of Ne^{20}
(6) Atom of Ne^{22}
(7) Atom of Na   O - Electron

Fig. 7. Aston's diagrams showing the number of electrons and protons inside various light nuclei (dark circles) and the "planetary electrons" (open circles) outside. (*Isotopes*, London: Arnold, 1922, p. 97; 2nd ed., 1924, p. 115.)

if electrons are comfortable inside the nucleus, it ought to be possible to inject some from the outside. Assisted by W. B. Kay, he tried the experiment, bombarding a mercury surface with 138–145 kilo-electron-volt electrons.[82] It is amusing to note that had the experiment succeeded, which it did not, Harkins would have produced a tiny supply of gold in Chicago.

THE EMERGENCE OF DIFFICULTIES, 1926–1931

The nuclear electron hypothesis, deeply entrenched in the minds of physicists,

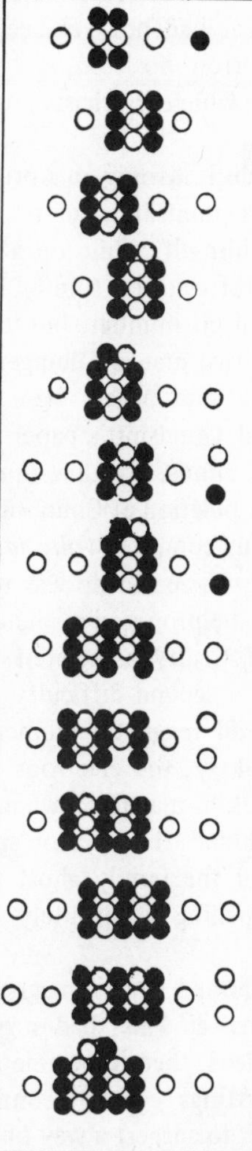

Number of diagram.		Atomic Number.	Nuclear Constitution.	Charge.	Mass.	Stability.	Description.
12		2	4+2 —	+1	5·007		Positively charged HeH
13		3	6+3 —	+1	6·0		Positively charged Li^6 atom
14		3	6+3 —	0	6·0	4·9*	Neutral Li^6 atom
15		3	7+4 —	+1	7·0		Positively charged Li^7 atom
16		3	7+4 —	0	7·0	4·9*	Neutral Li^7 atom
17		3	6+3 —	0	6·0(07)		Neutral Li^6H molecule
18		3	7+4 —	0	7·0(07)		Neutral Li^7H molecule
19		4	9+5 —	+1	9·0		Positively charged Be atom
20		4	9+5 —	0	9·0	3·3*	Neutral Be atom
21		5	10+5 —	+2	10·00		Doubly charged B^{10} atom
22		5	10+5 —	0	10·00		Positively charged B^{10} atom
23		5	10+5 —	0	10·00		Neutral B^{10} atom
24		5	11+6 —	+2	11·00		Doubly charged B^{11} atom

* Calculated from frequency of radiation.

Fig. 8. Aston's diagrams for the nuclei of various isotopes between $Z = 2$ and 5; The core or "nuclear constitution" of each is shown by the tightly packed protons (dark dots) and electrons (open circles), while the others are "thousands of times" more distant but still within the nucleus. (*Isotopes*, 1922 ed., p. 107; 1924 ed., p. 129.)

nevertheless became the focus of theoretical and experimental objections beginning in 1926. Widespread acceptance coexisted with fundamental contradictory evidence – a remarkable situation made possible by the absence of viable alternatives and an extreme reluctance, characteristic of the period, to develop theories based upon particles which had not yet been experimentally observed. Witness the cases of the neutron, positron, and neutrino. In what follows, we must be alert to the ways in which physicists rationalized heresy with belief.

The first to spot a flaw in the fabric was R. de L. Kronig, in April 1926,[83] following G. E. Uhlenbeck and S. Goudsmit's publication of the electron spin hypothesis. We now know that Kronig himself, while on a traveling fellowship in Europe, independently conceived of electron spin as a physical interpretation of Wolfgang Pauli's fourth quantum number, but that Pauli discouraged Kronig from publication when the two met in Tübingen in early 1925.[84] Werner Heisenberg, Hendrik Kramers, and others were similarly skeptical, and it was only after Uhlenbeck and Goudsmit's papers of 1925 and 1926 appeared[85] that they changed their minds. By that time Kronig had returned to the United States to take up a position at Columbia University, and as he later remarked, "In view of this complete *volte face* of the leading physicists . . . the only thing remaining for me to do was to call attention to all the difficulties still in the way of the proposed explanation."[86]

First there was the classical "pictorial" difficulty associated with the electron's relativistic "internal velocities"; but a second difficulty appeared to be much more serious to Kronig. Unless all magnetic moments "just happened to cancel," which seemed very unlikely, the electrons ought to contribute magnetic moments on the order of a Bohr magneton when forming "part of the nuclear structure." Thus, he concluded, the electron spin hypothesis "appears rather to effect the removal of the family ghost from the basement to the sub-basement, instead of expelling it definitely from the house."[87]

The point was that if hyperfine splitting stemmed from a magnetic coupling between the nucleus and orbital electrons, as Pauli had suggested in 1924,[88] and if there were electrons in the nucleus, then hyperfine splittings ought to be as large as ordinary Zeeman splittings – a flat contradiction with observation. O. W. Richardson was the first to suggest a way out. Noting that the tiny nucleus must be "a highly interlocked structure" containing "a very considerable number of electrons and protons" with little opportunity for "ordered spinning," he suggested that it was likely that "the electron gets rid of this angular momentum in the process of nucleus formation either by passing it on to the nucleus as a whole or else by radiating it away."[89] Richardson's rationale was repeatedly echoed in the literature as more and more accurate hyperfine measurements appeared, such as Goudsmit and E. Back's 1927 measurements on bismuth.[90]

In 1928 Kronig, now visiting the Physisch-Laboratorium der Rijks-Universiteit in Utrecht, noticed a second flaw in the fabric.[91] Ornstein and Van Wyk in that laboratory had just completed measurements on the band spectrum of the N_2^+ ion, and Kronig, citing F. Hund's theoretical studies,[92] concluded that these measurements indicated a spin of 1 (in units $h/2\pi$) for the nitrogen nucleus. "Since one thinks of the N nucleus . . . as built up of 14 protons and 7 electrons, a total of 21 particles," Kronig wrote, "one can be immediately astonished by this result." Thus, an odd number of spin-1/2 particles ought to produce a half-integer total spin and a total magnetic moment on the order of a Bohr magneton. "One is therefore probably required to assume," Kronig stated, "that in the nucleus the protons and electrons do not maintain their identity in the same way as in the case when they are outside the nucleus."[93] Even if one made the unlikely assumption that there exist several types of nitrogen nuclei of atomic weight 14, which might resolve the spin contradiction, the total magnetic moment of the nitrogen nucleus would still not agree with observation.

Even as Kronig was drawing these incisive conclusions, another chain of events had been set into motion which would make the contradictions still more acute. Enrico Fermi, determined to enter the emerging field of nuclear physics, had begun to send members of his group in Rome to laboratories throughout the world to learn appropriate experimental techniques. For the Rome group one natural bridge between the old and the new, between atomic and nuclear physics, was optical spectroscopy. Thus it happened that Franco Rasetti found himself at the California Institute of Technology in early 1929 on an International Education Board fellowship undertaking experiments on the Raman effect in diatomic gases, that is, measuring their rotational energy levels from their band spectra.[94]

Rasetti obtained strong, well-resolved Raman band spectra for diatomic oxygen and nitrogen using the mercury 2536-Ångstrom line as exciting radiation,[95] and his photographs displayed the evenly spaced lines he expected to find from the selection rules developed by Heisenberg, Hund, and Kronig. Moreover, whereas the O_2 lines were all of equal intensity, those of N_2 exhibited "the characteristic phenomena of alternating intensities." However, other experiments indicated that these N_2 lines alternated in intensity opposite to those of diatomic hydrogen — a point that was not clear previously, at least not to O. W. Richardson.[96] "Perhaps," Rasetti concluded, "it is significant for the properties of the nuclei that N_2 and H_2, which have a similar electronic structure . . . , behave in opposite ways as to the relative weight of odd and even rotational states."[97]

Rasetti's paper, which was published in the *Proceedings of the National Academy of Sciences*, immediately caught the attention of Walter Heitler and Gerhard Herzberg in Göttingen, who fired off an article to *Die Naturwissenschaften* liberally sprinkled with italics and headed by a bold title, "Does the Nitrogen Nucleus obey Bose Statistics?"[98] They concluded that it does, basing their argument on a recent proof by E. P. Wigner and E. E. Witmer,[99] resting solely on symmetry considerations, that the total eigenfunction of a symmetric molecule must be even for even rotational states, odd for odd, and that symmetric *spin* eigenfunctions always have a larger statistical weight than antisymmetric ones. The measurements reported by Rasetti meant, therefore, that the total eigenfunction for H_2 is always antisymmetric in the nuclei, for N_2 always symmetric. "The latter means," they concluded, "that the *N-nuclei satisfy Bose statistics*."[100] They observed in a footnote that this same conclusion did *not* follow from Ornstein and Van Wyk's measurements, since the N_2^+ ion is *not* a symmetric molecular system.[101]

This result, Heitler and Herzberg stated, is "extraordinarily surprising." For Wigner had also rigorously proved, in a paper still in press in the *Proceedings of the Hungarian Academy*,[102] that a composite system consisting of an odd number of spin-1/2 particles must obey Fermi statistics, an even number Bose statistics. If correct, therefore, Rasetti's measurements meant that Wigner's rule was "*no longer valid in the nucleus*." "[It] seems as if *the electron in the nucleus, together with its spin also loses its ability to determine the statistics of the nucleus* . . . ." "It is only certain that they retain their charge." "Naturally," they concluded, "the correctness of this consideration stands and falls with the correctness of Rasetti's analysis of the Raman spectrum of N_2."[103]

The seriousness of this conflict was again emphasized at a conference on molecular spectra and molecular structure at the University of Bristol in September 1929, both by W. E. Garner and J. E. Lennard-Jones in their introductory remarks, and by Robert S. Mulliken of the University of Chicago in his extensive paper.[104] "It would seem," Mulliken concluded, "either that there is an error in Rasetti's measurements," or that the symmetry requirement "is incorrect for molecules with complex nuclei, or that something has been overlooked in its application here."[105] The following month H. Schüler and H. Brück also cast doubt on Rasetti's results by suggesting that his observations may have been invalidated by diffraction effects.[106]

These demurrers, as well as the obvious high importance of the measurements, stimulated Rasetti to repeat his experiments using even higher resolution instrumentation on his return to Rome. The new results, which he reported on March 10, 1930, completely confirmed his earlier work.[107] The beautiful Raman band spectra he obtained for O_2 and N_2, as shown in Figure 9, simply left no doubt as to the correctness of his observations.

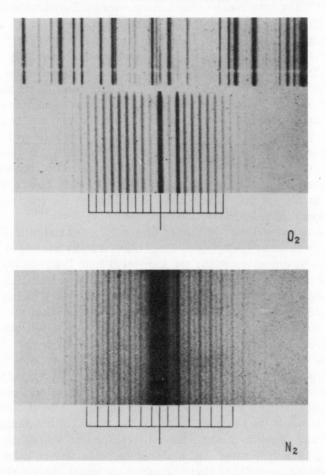

Fig. 9. Rasetti's Raman band spectra for O_2 and N_2. Proper interpretation shows that both the oxygen and nitrogen nuclei obey Bose statistics, in contrast to the hydrogen nucleus, which obeys Fermi statistics. ("Über die Rotations-Ramanspektren von Stickstoff und Sauerstoff," *Z. Phys.*, 61 (1930), 600. Reproduced by permission of Springer-Verlag.)

The contradictory behavior of the nuclear electrons regarding their spin and statistics, with particular reference to the nitrogen nucleus, was widely appreciated. In February 1930 Guido Beck in Leipzig, during the course of an extended study on the systematics of isotopes, emphasized it,[108] and J. Dorfmann at the Physico-Technical Institute of the University of Leningrad noted that it was "consequential for the entire future theory of atomic nuclei."[109] The latter case is particularly noteworthy. Dorfmann perceptively analyzed existing data on magnetic susceptibilities, concluding that "nuclei which contain 2, 3, 4, 5, 6, or 7 electrons have no magnetic moments on the order of 1 M_B [Bohr magneton]." This result agreed "completely with the spectroscopic data," including those for the radioactive element radium, suggesting its generality. To Dorfmann there were only three explanations possible: that the nuclear electron spins were "always somehow mutually compensated," that the electrons lost their spins "as soon as they enter the nucleus," as suggested by Heitler and Herzberg, or that "there are absolutely no electrons in the nucleus." Regarding the last possibility, Dorfmann called attention to the efforts of D. Iwanenko and V. Ambarzumian, then underway, to construct a theory of beta decay based upon P. A. M. Dirac's relativistic wave equation for the electron and the assumption that electrons inside nuclei "lose their individuality" just as photons do in atoms.[110]

The spin and statistics difficulties were still further emphasized in 1931 by Paul Ehrenfest and J. R. Oppenheimer who, evidently unaware of Wigner's earlier work, proved essentially the same result from Pauli's exclusion principle for a symmetric molecular system.[111] P. Güttinger and Pauli pointed out at about the same time that the light isotope of lithium, Li^6, apparently should join nitrogen as a problematic element, since they had found strong indications that the spin of Li^6 was zero, while it evidently contained an odd number of nuclear constituents (6 protons plus 3 electrons).[112] A most insightful solution to the difficulty was suggested in July 1931 by R. M. Langer, who pointed out that according to Dirac's theory the effective magnetic moment of an electron approaches zero as its velocity approaches the velocity of light, and hence "inside the nucleus . . . the electron would no longer behave as a magnet."[113] J. H. Van Vleck, in his classic treatise *Electric and Magnetic Susceptibilities* (whose preface was signed in June 1931), poignantly pointed out in a footnote what was at stake: "If the spin magnetic moments of electrons within the nuclei did not very nearly compensate each other, our whole theory of susceptibilities would be upset. . . ."[114]

The difficulties besetting the nuclear electron hypothesis did not end with considerations of spin and statistics. Following Heisenberg's discovery of the uncertainty principle in 1927 and the publicity Bohr gave it, together with his own principle of complementarity, in the fall of 1927 at the Volta and Solvay Conferences, it could be demonstrated easily that an electron confined to nuclear dimensions would have to possess momentum, and hence energy, far in excess of known nuclear binding energies. Remarkably, however, this simple argument seems to be virtually absent in the literature between 1927 and 1930 — a reticence quite likely stemming from the unfamiliarity, even radicalness, of Heisenberg's and Bohr's principles, and the fact that while the first barrage in the Bohr-Einstein debate was fired at the 1927 Solvay Conference, its depth was not plumbed until the 1930 Solvay Conference.[115] Perhaps a remark of Rutherford's in early 1929, that "it is not easy to confine an electron in the same cage with an α-particle," was an allusion to the uncertainty principle argument.[116] More likely, however, it was an allusion to a related difficulty which had just became known and which, by contrast, was widely discussed in private and in print — the so-called Klein paradox.

This paradox, discovered by Oscar Klein toward the end of 1928 while on an extended visit to Bohr's Institute in Copenhagen, may be formulated as follows: According to Dirac's wave equation for the electron, it follows that an electron of sufficiently high energy incident upon a high and steeply rising potential barrier, instead of being reflected from the barrier, has a high probability of escaping through it by being transformed into a particle of negative mass, that is, by simply being transformed from a $+m$ particle into a $-m$ particle. From these considerations it follows that an electron can never be contained within a nuclear potential well. How, then, could there be electrons inside the nucleus?[117]

The clarity and precision of this paradox was immediately appreciated, and it soon cropped up in many places. Gamow investigated it in Cambridge at the end of 1929 and beginning of 1930;[118] Bohr debated its meaning with Dirac by letter in November and December 1929;[119] J. Kudar, loosely connected to Schrödinger in Berlin, attempted a resolution of it;[120] F. G. Houtermans discussed it in an review article in 1930;[121] and Gregory Breit and his colleagues at the Carnegie Institution in Washington regarded it as one of the "main unsolved questions at present"[122] — to cite but some of the references to it that I have located.

In one way or another, of course, all of the difficulties associated with the nuclear electron hypothesis impinged upon contemporary attempts to construct a satisfactory theory of beta decay, especially following Ellis and Wooster's 1927 experiments. Joan Bromberg has described how Bohr, beginning in mid-1929 and stimulated by G. P. Thomson's 1928–1929 attempt to interpret the continuous beta-ray spectrum along wave-mechanical lines, became convinced that energy conservation is violated in beta decay.[123] She has shown how Bohr's conviction must be seen within the context of his increasing expectation that beta decay and the behavior of nuclear electrons lay outside "existing quantum physics" and hence demanded the development of a "new physics."[124] Bohr first presented his ideas orally, emphasizing the "remarkable 'passivity' of the intra-nuclear electrons," in May 1930 in London and Cambridge (Faraday and Scott lectures) and in October 1931 in Rome. He wrote them up for publication in early 1932.[125] Well before their oral and written presentations, however, Bohr's views became widely known. He mentioned them in a letter to N. F. Mott in Cambridge on October 1, 1929, and Mott obviously reported them to Rutherford, for on November 19 Rutherford conveyed his reaction to Bohr by letter in the following terms: "I have heard rumours that you are on the war path and wanting to upset Conservation of Energy, both microscopically and macroscopically. I will wait and see before expressing an opinion, but I always feel 'there are more things in Heaven and Earth than are dreamed of in our Philosophy.'"[126]

Pauli's earlier reaction, in July 1929, as described by Bromberg, was even more strongly negative, as was Dirac's in December.[127] Heisenberg, however, willingly relinquished conservation of energy, momentum, and even charge within a well-defined lattice world (*Gitterwelt*) and unsuccessfully attempted to devise a model of the nucleus consisting "only of protons and (slow) light quanta of mass M, not of electrons."[128] Similarly, Kudar incorporated Bohr's ideas into his own theorizing on beta decay following a letter from Bohr on January 28, 1930,[129] and Houtermans, in the 1930 review article already cited above, discussed Bohr's ideas as well.[130]

It was, of course, in this uncertain atmosphere that Pauli, implacably opposed to Bohr's program, proposed the existence of a "magnetic neutron" to resolve, he hoped, all of the outstanding difficulties at a single stroke.[131] Unwilling to forego a holiday ball in Zurich for a physics conference in Tübingen, Pauli first outlined his hypothesis in a letter to the conference participants, particularly Geiger and Meitner, on December 4, 1930. There is, Pauli wrote,

... the possibility that there could exist in the nuclei electrically neutral particles that I wish to call neutrons, which have spin 1/2 and obey the exclusion principle, and additionally differ from light quanta in that they do not travel with the velocity of light: The mass of the neutron must be of the same order of magnitude as the electron mass and, in any case, not larger than 0.01 proton mass. – The continuous β-spectrum would then become understandable by the assumption that in β decay a neutron is emitted together with the electron, in such a way that the sum of the energies of neutron and electron is constant.[132]

Pauli went on to predict that such a neutron would behave like a magnetic dipole at rest (moment $\mu < e \cdot 10^{-13}$ cm), and would have about ten times the penetrating power of a gamma ray. He discussed this hypothesis again in Pasadena in June 1931, in Ann Arbor a few weeks later, and yet again in Rome in October 1931, especially with Fermi and with Bohr who, as has already been mentioned, presented his opposing point of view. It is important to realize, however, that even at that late date Pauli regarded his magnetic neutron as a fundamental constituent of the nucleus, and those who learned about it accepted it on those terms.[133]

One insightful way of monitoring and summarizing the changing attitude toward the nuclear electron hypothesis between 1928 and 1931 is by considering points in the intellectual trajectory of George Gamow, the physicist who through his quantum theory of alpha decay made no doubt the most important theoretical contribution to nuclear physics of the period.[134] Gamow has described how as a visitor in Göttingen in the summer of 1928 he went into the university library, came across Rutherford's 1927 paper in the *Philosophical Magazine* describing *inter alia* the uranium paradox, and immediately understood that in radioactive alpha decay he was confronted with a typical quantum-mechanical tunneling phenomenon.[135] Working out the problem in detail, he provided the first satisfactory explanation of the well-known Geiger-Nuttall relationship between decay constant and alpha energy, a major achievement and striking confirmation of his theory. He submitted his paper for publication on August 2, 1928, and shortly thereafter left Göttingen and Born for Copenhagen and Bohr. Virtually simultaneously, in November 1928, R. W. Gurney and E. U. Condon at Princeton University arrived independently at the same theory.[136] It is particularly noteworthy for our story, however, that while Gamow avoided entirely the problem of beta decay, Gurney and Condon, accepting the nucleus as the "habitat of the nuclear electrons," treated beta decay as a tunneling effect as well[137] – an approach that was possible only because *they* avoided entirely the problem of gamma radiation.

Gamow devoted the next several years to attacking problems in nuclear physics, extending his theory of alpha decay,[138] considering the inverse problem of alpha bombardment with crucial consequences for the work of J. D. Cockcroft and E. T. S. Walton,[139] advancing the liquid-drop model of the nucleus,[140] and exploring a host of related problems. Between 1928 and 1930 we find him firmly committed to the nuclear electron hypothesis,[141] accepting Bohr's view on nonconservation of energy in beta decay,[142] but at the same time being deeply concerned with the Klein paradox and other difficulties of nuclear theory and quantum electrodynamics.[143] By May 1931, when he signed the preface to his remarkable first book, *Constitution of Atomic Nuclei and Radioactivity,*[144] the tensions and contradictions had become crystal clear to him.

Thus, the most striking characteristic of his book was that while treating all nuclear properties and phenomena from the point of view of the nuclear electron hypothesis, at every point where that hypothesis entered he included a paragraph or paragraphs set off by special symbols – skull and cross bones in the manuscript,[145] transformed into "lazy Ss" by Oxford University Press – explaining its contradictions and difficulties. Side-by-side with the accepted picture of nuclei as "built up of elementary particles – protons and electrons,"[146] we find beautifully clear presentations of the uncertainty principle difficulty, the Klein paradox, the spin and statistics difficulties (with special reference to the case of nitrogen), the apparent loss of the electron's magnetic moment inside the nucleus, and the difficulties of beta decay.[147] I will quote but a single summarizing sentence: "The usual ideas of quantum mechanics absolutely fail in describing the behaviour of nuclear electrons; it seems that they may not even be treated as individual particles . . . , and also the concept of energy seems to lose its meaning . . . . "[148]

The influence of Bohr on Gamow is readily apparent here. Many years later Léon Rosenfeld, similarily influenced, reflected on the entire state of affairs as he experienced it circa 1931:

[We] were extraordinarily light-hearted about those well-known difficulties, not because we saw any simple way out, but because there were so many of them. The "intranuclear" electrons had so strange properties that we hoped that all those difficulties would some-how cancel each other and give us a beautiful theory. But we were far from guessing how it would come about.[149]

CHADWICK'S DISCOVERY OF THE NEUTRON, 1932

"But we were far from guessing how it would come about." The enormous effort expended throughout the 1920s in constructing nuclear models firmly based on the nuclear electron hypothesis directly conflicted with the fundamental difficulties that emerged after 1926 in regard to that same hypothesis. Nothing displays the value of experimental physics more than the flat contradiction in these theoretical programs; nothing better reveals the crucial importance of experimentalists addressing Nature on their own terms. For in the event it was Chadwick's experimental discovery of the neutron in 1932 that fundamentally reoriented nuclear theory and pointed the way toward the resolution of its past contradictions and paradoxes.

Although there are a number of references to the neutron in the literature prior to 1932,[150] and although Harkins adamantly claimed its discovery as his own after the fact,[151] these speculations and claims suffered from serious deficiencies and can be discredited. We can, therefore, safely reaffirm the traditional account, which begins at the end of January 1932 when an issue of the *Comptes rendus* arrived at the Cavendish Laboratory in Cambridge containing Irène Curie and Frédéric Joliot's conclusion, which agreed with W. Bothe and R. Becker's earlier analysis, that polonium alpha particles striking beryllium produce a highly energetic gamma radiation which in turn ejects protons from paraffin or other hydrogenous substances by means of a Compton scattering process.[152] Rutherford's outburst, "I don't believe it,"[153] agreed with Chadwick's reaction to that interpretation. Chadwick therefore carried out "many experiments,"[154] analyzed them on the basis of the conservation laws, and thereby obtained strong evidence that neutrons, not gamma rays, were being produced by the polonium alpha particles according to the reaction

$$_2^4 \text{He} + {}_4^9 \text{Be} \rightarrow {}_6^{12}\text{C} + {}_0^1 n$$

On February 17 he sent off a preliminary note to *Nature* reporting his discovery;[155] on February 23 he discussed it at a meeting of the Kapitza Club in Cambridge;[156] the following day he described it to Bohr by letter;[157] on April 28 he discussed it at a meeting of the Royal Society;[158] and on May 10 he sent off a full report to the Royal Society's *Proceedings.*[159]

The principal point of interest to us is Chadwick's precise conception of the neutron, the particle he had just discovered. He did not address this point in his note to *Nature* in February, but he left no doubt on it in his remarks to the Royal Society in April and in his full report in May. We there find a powerful example of the influence of tradition in physics, for Chadwick's neutron turns out to be identical to Rutherford's neutron of a dozen years earlier. Thus, to Chadwick the "simplest hypothesis" (his characterization) was that the neutron "consists of a proton and an electron in close combination, giving a net charge 0 and a mass which should be slightly less than the mass of the hydrogen atom."[160]

44   ROGER H. STUEWER

This point could be demonstrated by direct calculation, but not on the basis of the beryllium experiments, because the mass of beryllium had not yet been measured. Rather, Chadwick had to appeal to his similar experiments with boron, for which the mass-energy balance could be expressed by the equation

$$^{11}_{5}B + ^{4}_{2}He + W_{\alpha} = ^{14}_{7}N + ^{1}_{0}n + W_N + W_n$$

In this equation the masses of boron, helium, and nitrogen were known from Aston's measurements (11.00825 ± 0.0016, 4.00106 ± 0.0006, 14.0042 ± 0.0028, respectively), and the kinetic energies of the polonium alpha particles, nitrogen recoil nucleus, and neutron were determined experimentally (0.00565, 0.00061, and 0.0035 mass units). By simple subtraction, Chadwick found the mass of the neutron to be 1.0067 mass units, though taking account of the experimental errors widened the range from 1.005 to 1.008 mass units.

"Such a value," Chadwick observed, "is to be expected if the neutron consists of a proton and an electron," because the sum of the proton and electron masses was 1.0078 mass units, which again by subtraction gave as the mass defect or binding energy of the proton-electron combination the value 1–2 million electron volts. This, Chadwick noted, "is quite a reasonable value. We may suppose that the proton and electron form a small dipole, or we may take the more attractive picture of a proton embedded in an electron. On either view, we may expect the 'radius' of the neutron to be a few times 10^{-13} cm."[161]

Now, such a neutron, while electrically neutral, would nevertheless possess an "extremely small" but finite electric field at "distances on the order of 10^{-12} cm." Hence, it would interact electrically with matter, and Chadwick in fact sketched a possible shape for a neutron-nucleus interaction potential, as shown in Figure 10. He also observed, in a remarkable sentence, that neutron-proton and neutron-electron collision experiments would be of "peculiar interest": "A detailed study of these collisions with an elementary particle is of special interest, for it should provide information about the structure and field of the neutron, whereas the other collisions will depend mainly on the structure of the atomic nuclei."[162] Chadwick noted that such experiments in fact were already underway at the Cavendish, and those on electrons were in accord with the low collision probabilities anticipated by Bohr and H. S. W. Massey.[163]

Along with a convincing interpretation of the Paris and Cambridge experiments, therefore, Chadwick saw his discovery of the neutron fundamentally

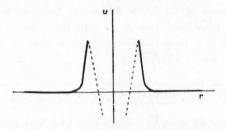

Fig. 10. Chadwick's diagram of the interaction potential U versus separation distance r for a neutron-nucleus collision. The sharp increase in potential occurs at a distance only slightly greater than the nuclear radius, and is relatively low to permit the neutron to enter easily. ("Existence of a Neutron," *Proc. Roy. Soc.,* **A 136** (1932), 702. Reproduced by permission of the Royal Society.)

as vindicating Rutherford's 1920 prediction, and specific model, rather than as resolving the contradictions associated with the nuclear electron hypothesis. It is true that in his general remarks at the end of his full report he noted that there would be "certain advantages" in avoiding the presence of "uncombined electrons" by constructing nuclei out of alpha particles, neutrons, and protons only, but at the same time he resolutely reiterated his belief that the "simplest assumption" was that the neutron was a "complex particle." And nothing could reveal his fundamental bias more clearly than his statement that viewing the neutron as "an elementary particle . . . has little to recommend it at present, except the possibility of explaining the statistics of such nuclei as N^{14}."[164]

In fact, to fully appreciate the depth of Chadwick's conviction, we need only refer to the extensive discussion that took place at the Royal Society about two weeks before Chadwick submitted his full report.[165] At that discussion, Chadwick's discovery played center stage; Rutherford, C. D. Ellis, and N. F. Mott surveyed other recent developments in nuclear physics; and R. H. Fowler, J. C. McLennan, and F. A. Lindemann, in particular, aired at length all of the difficulties associated with the nuclear electron hypothesis. This discussion evidently did not cause Chadwick to change his point of view. Indeed, even a full year later, on May 25, 1933, when Chadwick delivered his Bakerian Lecture, "The Neutron," before the Royal Society, his fundamental bias was still apparent.[166] The mass of the neutron, he again emphasized, is "consistent with the view that the neutron consists of a proton and an electron." At the same time it is clear that by then he had become much more impressed with the spin and statistics difficulties. Thus he concluded by straddling the fence: "It seems necessary for the present to recognize these difficulties and, while retaining the hypothesis that the neutron is complex for some purposes, to regard it as an elementary unit in the structure of atomic nuclei."[167]

THE NEUTRON IN NUCLEAR THEORY, 1932–1933

As we now turn to the broader influence of Chadwick's discovery, it seems unquestionably significant that the first person to state unequivocally that the "chief point of interest is how far the neutrons can be considered as elementary particles (something like protons or electrons)" was someone far removed from Rutherford and the Cavendish tradition, namely, D. Iwanenko of the Physico-Technical Institute of the University of Leningrad.[168] We have already noted how in 1930 J. Dorfmann, working with Iwanenko in the same institute, exhibited great independence of thought by explicitly considering the possibility of expelling electrons entirely from the nucleus.[169] Now, on April 21, 1932, two months after the appearance of Chadwick's note in *Nature*, Iwanenko sent a note to the same journal expressing the exceptional point of view above. Four months later, in a paper published in the *Comptes rendus*, he provided a more detailed basis for constructing nuclei out of protons and neutrons only, treating neutrons as spin-1/2 elementary particles obeying Fermi statistics, and recalling his and Ambarzumian's earlier idea that beta emission is analogous to the emission of photons by atoms.[170]

The crucial distinction between regarding the neutron as elementary or complex in nuclear theory was pointed out in a review article by H. Kallmann and H. Schüler in 1932, evidently completed shortly after they learned of Chadwick's discovery. For throughout most of their long discussion they assumed that nuclei are "built up of protons and electrons,"[171] but at one point, in a passage that beautifully displays the confusion of the period, they observed that an "entirely different standpoint" was possible:

Thus the nucleus should not only consist of protons and electrons, but at least also of a third type of elementary constitutent particle, namely neutrons (Pauli); a hypothesis which has become probable recently through the discovery of neutrons in atomic disintegration .... We would like to stress ... that one only achieves an "explanation" of the [spin and other] discrepancies if one regards the neutrons as new elementary particles. If they are constructed out of a proton and electron, however, the above difficulties would remain undiminished.[172]

Blended together are Pauli's magnetic neutron and Chadwick's neutron; clearly distinguished is the crucial significance of regarding the neutron as elementary or complex: *A compound neutron entailed all of the past difficulties of the nuclear electron hypothesis.*

This distinction was missed by F. Perrin in Paris, who on April 18, 1932, stimulated by Chadwick's discovery, examined the possible structure of various nuclei based upon G. Fournier's earlier use of a compound neutron.[173] It was also missed by Fournier himself one week later and by Marie Curie in July.[174] That it escaped Bohr, who embraced the complex neutron following Chadwick's discovery, and indeed "left the difficulties of nuclear structure almost exactly where they were before," has already been pointed out by Joan Bromberg.[175] Bromberg and others also have already emphasized that this distinction was blurred throughout by Heisenberg in his seminal three-part paper of 1932 — that, in fact, in all three parts a tension is present between the neutron as elementary particle and as complex particle.[176]

Thus in Part I, written in June, Heisenberg, referring to Iwanenko, notes that the neutron as an elementary particle would resolve the spin and statistics difficulties, whereas as a complex particle it could decay inside the nucleus into a proton and electron and hence account for beta decay as well as explain certain scattering experiments.[177] As an elementary particle, Heisenberg's p coordinate takes on only two values, $+1$ for the neutron state, -1 for the proton state; as a complex particle the neutron is bound to the proton by exchanging its constituent electron, analogous to the exchange of the electron in the chemical bonding of the H_2^+ ion.[178]

In Part II, written in late July, Heisenberg again emphasizes the advantages of conceiving the neutron as a "solid elementary unit"; yet, from its mass defect or binding energy, on the order of 1 million electron volts according to Chadwick, one is led to view it as if "composed of a proton and an electron."[179] The latter picture, Heisenberg now points out, entails precisely the uncertainty principle difficulty. For if confined within the neutron, whose diameter Heisenberg takes to be on the order of the classical radius e^2/mc^2, the electron's uncertainty in momentum, Δp, would be on the order of

$$\Delta p \sim \frac{h}{2\pi\Delta q} \sim \frac{h}{2\pi}\frac{mc^2}{e^2} = \frac{hc}{2\pi e^2}\,mc$$

corresponding to an energy E on the order of

$$E \sim c\Delta p = \frac{hc}{2\pi e^2}\,mc^2 \approx 137\,mc^2$$

This value is two orders of magnitude greater than the electron's rest-mass energy. Heisenberg showed that it also was in rough agreement with the resonant energy of a classical oscillator having an effective cross section as large as that of a neutron, as estimated from certain gamma-ray scattering data. The above figure meant that a "unique definition of the concept 'binding energy' is impossible for the electron in the neutron on account of the failure of the energy law in β-decay." "In some respects," he summarized, "the neutron behaves as a quantum mechanical system of very large binding energy [i.e., as an elementary particle]; in contrast, its mass defect is very small."[180]

In Part III, written in December, Heisenberg included a long final section entitled "Discussion of the Assumptions on the Nature of the Neutron."[181] He there expressed doubt that the two pictures, of the elementary and the complex neutron, could be joined together in a single consistent model. It will come as no surprise to learn that in the fall of 1932 Harkins delivered a paper at the National Academy of Sciences whose clear intent was to claim priority once again by reiterating his "prediction" of the complex neutron and displaying its influence on Heisenberg.[182]

The relative ease with which physicists separated from the Cambridge and Copenhagen traditions could embrace the neutron as a new elementary particle is illustrated, finally, by the case of Ettore Majorana, working in Fermi's group in Rome. According to Segrè, immediately upon learning of the Paris experiments Majorana burst out: "Oh, look at the idiots; they have discovered the neutral proton, and they don't even recognize it."[183] Majorana's conviction was no doubt conditioned in part by discussions in Rome on Rasetti's Raman band spectra experiments, which as we have seen were so important in focusing attention on the spin and statistics difficulties of nitrogen.[184]

Segrè recalls that Majorana actually began to develop a neutron-proton theory of nuclear structure shortly before Fermi was to leave for Paris to attend the Congrès International d'Électricité, which took place July 5–7, 1932.[185] Fermi urged Majorana to publish his ideas, but – brilliant eccentric that he was – Majorana refused, considering them incomplete, and even going so far as to make it impossible for Fermi to publicize them at the Paris conference. Thus we find that Fermi, in his Paris paper, simply surveyed the present state of nuclear physics, presented the electron-proton model along with its difficulties, referred along the way to Pauli's "neutron" hypothesis (Fermi clearly had not yet adopted the term "neutrino"), and appended only two short paragraphs at the end discussing Chadwick's discovery of a new type of "neutron" and Cockcroft and Walton's recent experiments.[186] Fermi evidently was just on the verge of revising his ideas on nuclear structure.

Majorana ultimately published his theory of the nucleus – in which the neutron and proton are treated as distinct elementary particles – as a result of Heisenberg's urging while Majorana was visiting Heisenberg in Leipzig in March 1933.[187] Following Hans A. Bethe's recent summary, Majorana's theory was "completely modern" in that he discarded the chemical bonding analogy and instead started from "the observed properties of nuclei, particularly the saturation of nuclear forces and the particular stability of the α-particle."[188] He realized that he could achieve saturation by an exchange force, so he postulated one, but one different from Heisenberg's: instead of Heisenberg's picture of a neutron and proton located at two different positions exchanging only their charges, with their spins remaining at the original positions, Majorana's eliminated Heisenberg's ρ coordinate and involved the exchange of spin as well. In this way Majorana found that the forces saturated at $^4_2$He, the alpha particle, while Heisenberg's would have saturated at the deuteron. The latter particle, as we shall see, was found to be not particularly stable.

THE DECLINE OF THE NUCLEAR ELECTRON HYPOTHESIS, 1933–1934

The theories of nuclear structure that were developed following Chadwick's discovery of the neutron indicated the presence of entirely new, nonelectromagnetic forces inside the nucleus, and hence were fundamental to the future of nuclear physics. These developments have been described recently by Sir Rudolf Peierls,[189] and lie beyond the scope of my paper. Central to it, however, is the question of the general acceptance of the neutron as a new elementary particle rather than as an electron-proton compound which entailed, knowingly or not, all of the basic difficulties of the venerable nuclear electron hypothesis.

While I have noted the exceptional cases of Iwanenko and Majorana, it is possible to point to still further evidence that the complex-neutron model enjoyed widespread appeal in the months following Chadwick's discovery. I refer here to a lecture delivered by Merle A. Tuve at the Franklin Institute in Philadelphia on February 2, 1933, in which Tuve described his and his colleagues' attempts, begun in 1926 under the direction of Gregory Breit, to construct an accelerator at the Cargnegie Institution in Washington for exploring the structure of nuclei.[190] Tuve illustrated his lecture with, for example, a "tentative table" of light nuclei and a picture of the disintegration of aluminum, showing neutrons as electron-proton compounds, as seen in Figure 11.

THE STRUCTURE OF THE NUCLEUS OF AN ATOM DETERMINES ITS
MASS AND ITS POSITIVE CHARGE, HENCE DETERMINES WHAT KIND
OF A CHEMICAL ELEMENT IT FORMS

TENTATIVE TABLE OF ATOMIC NUCLEI

ATOMIC NUMBER	CHEMICAL ELEMENT	MASS	NUCLEAR STRUCTURE		
1	HYDROGEN	1		1 PROTON	
1	HYDROGEN₂	2		1 PROTON, 1 NEUTRON	
2	HELIUM	4		2 PROTONS, 2 NEUTRONS	1 ALPHA-PARTICLE
3	LITHIUM₆	6		3 PROTONS, 3 NEUTRONS	1 ALPHA-PARTICLE, 1 PROTON, 1 NEUTRON
3	LITHIUM₇	7		3 PROTONS, 4 NEUTRONS	1 ALPHA-PARTICLE, 1 PROTON, 2 NEUTRONS
4	BERYLLIUM	9		4 PROTONS, 5 NEUTRONS	2 ALPHA-PARTICLES, 1 NEUTRON

THIS TABLE IS IN PROCESS OF BEING CONSTRUCTED AND ESTABLISHED

DISINTEGRATION OF ALUMINUM NUCLEUS WHEN STRUCK
BY ALPHA-PARTICLE FROM RADIUM

NOTE: A NUCLEAR COLLISION IS SIMPLY THE INTERACTION
OF ELECTRIC AND MAGNETIC FORCES

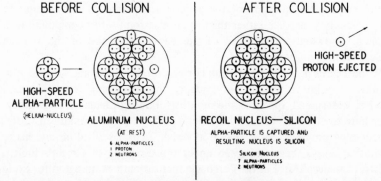

BEFORE COLLISION   AFTER COLLISION

HIGH-SPEED ALPHA-PARTICLE (HELIUM-NUCLEUS)

ALUMINUM NUCLEUS (AT REST)
6 ALPHA-PARTICLES
1 PROTON
2 NEUTRONS

HIGH-SPEED PROTON EJECTED

RECOIL NUCLEUS—SILICON
ALPHA-PARTICLE IS CAPTURED AND RESULTING NUCLEUS IS SILICON
SILICON NUCLEUS
7 ALPHA-PARTICLES
2 NEUTRONS

Fig. 11. Tuve's "tentative table" of light nuclei, and his picture of the disintegration of aluminum by alpha particles, showing neutrons as electron-proton compounds. ("The Atomic Nucleus and High Voltages," *J. Franklin Inst.*, 216 (1933), 5, 8. Reproduced by permission of the Franklin Institute.)

Tuve no doubt would have agreed both with K. T. Bainbridge's remark, that the "terms electron and proton must be used advisedly," and with K. K. Darrow's, that it "would be more elegant to design nucleus models out of two fundamental particles only, say the proton and the negative electron .... "[191] The naturalness and appeal of the complex neutron in early to mid-1933 can hardly be disputed. I have already pointed out Chadwick's own attitude on the matter in his Bakerian Lecture of May of that year.

It is impossible to monitor changing attitudes precisely, but it seems clear that the complex neutron lost its viability in the eyes of Chadwick and most of his contemporaries between October 1933 and October 1934, a year that began and ended with landmark conferences on nuclear physics, the Septième Conseil de Physique Solvay, held in Brussels, October 22–29, 1933, and the International Conference on Physics, held in London, the first week in October 1934.[192] The first saw the emergence of fundamental problems whose resolution was achieved by the time of the second.

Each of the six papers delivered at the seventh Solvay Conference – by Cockcroft, Chadwick, the Joliot-Curies, Dirac, Gamow, Heisenberg – deserves attention, but I shall concentrate on only two, Chadwick's and the Joliot-Curies'. The former, particularly the section entitled "The Neutron," reveals that Chadwick was still inclined to straddle the fence on the nature of the neutron.[193] In favor of its complexity was its mass, which he now calculated from two reactions:

$$^{11}_{5}B + ^{4}_{2}He \rightarrow ^{14}_{7}N + ^{1}_{0}n \quad \text{and} \quad ^{7}_{3}Li + ^{4}_{2}He \rightarrow ^{10}_{5}B + ^{1}_{0}n$$

using Aston's values for the masses of the boron and nitrogen isotopes, and Bainbridge's and Cockcroft's for lithium. These values in the first reaction and those of the particle kinetic energies differed only in the fourth decimal place, if at all, from those he had used earlier. Then, taking account of the second reaction and the experimental errors in all of the measurements, Chadwick concluded that the maximum mass of the neutron was 1.0072 ± 0.0005 mass units. The minimum, which he calculated from Bainbridge's measurements of the deuton (now deuteron) mass, was 1.0056 mass units. These values, Chadwick noted, were consistent with the "conception that the neutron is a complex particle" with a mass defect or binding energy of 1 million electron volts.[194]

This point, however, was "not decisive." Arguing against this picture were several factors: that one could show that the probability for decomposition into a proton and electron in a collision process was extremely low;

that there were quantum-mechanical, spin, and statistics difficulties; that the ordinary hydrogen atom — another proton-electron system — obviously did not collapse spontaneously into a neutron with emission of energy; and that the proton, contrary to what one might expect, did not appear to be composed of a neutron and the recently discovered positive electron. All of the above observations suggested that the neutron was an elementary particle. However, once again arguing the contrary position, Chadwick explained at great length how certain experiments might be accounted for on the assumption that either the neutron or the proton was a complex particle.[195]

Chadwick's paper drew no searching comments during the discussion; the true challenge emerged during the Joliot-Curies' paper, on the penetrating radiation emitted by atoms under alpha bombardment.[196] For the Joliot-Curies pointed out, politely but unambiguously, that Chadwick's calculation of the neutron mass rested on the assumption that none of the reaction energy was carried away by gamma rays, and they then went on to report a host of experiments on light elements in which this occurred. They also displayed cloud chamber photographs in which gamma rays produced elec-tron-position pairs by direct materialization of energy, and they suggested that positron emission opened up the possibility of new reactions occurring. For example, alpha particles incident on boron could yield, in addition to carbon, *either* a proton *or* a neutron and positron, according to the reactions

$$^{10}_{5}B + ^{4}_{2}He \rightarrow ^{13}_{6}C + ^{1}_{1}H$$

$$^{10}_{5}B + ^{4}_{2}He \rightarrow ^{13}_{6}C + ^{1}_{0}n + e^{+}$$

Similarly, alpha particles incident on aluminum could lead to either of the following two reactions:

$$^{27}_{13}Al + ^{4}_{2}He \rightarrow ^{30}_{14}Si + ^{1}_{1}H$$

$$^{27}_{13}Al + ^{4}_{2}He \rightarrow ^{30}_{14}Si + ^{1}_{0}n + e^{+}$$

Finally, alpha particles incident on beryllium could produce either a gamma ray or an electron-positron pair, according to the reactions

$$^{9}_{4}Be + ^{4}_{2}He \rightarrow ^{12}_{6}C + ^{1}_{0}n + h\nu$$

$$^{9}_{4}Be + ^{4}_{2}He \rightarrow ^{12}_{6}C + ^{1}_{0}n + (e^{+} + e^{-})$$

All of which served as background to the climax of the Joliot-Curies' paper, a "new basis for the calculation of the mass of the neutron."[197]

Focusing on the first pair of reactions above, the mass-energy balances became

$$^{10}_{5}B + {}^{4}_{2}He + W_{\alpha} = {}^{13}_{6}C + {}^{1}_{0}n + e^{+} + W_{1}$$

$$^{10}_{5}B + {}^{4}_{2}He + W_{\alpha} = {}^{13}_{6}C + {}^{1}_{1}H + W_{2}$$

where W_1 is the combined kinetic energy of the recoil carbon nucleus, the neutron, and the positron, and W_2 that of the recoil carbon nucleus and proton. By simple subtraction one obtained

$$W_2 - W_1 = {}^{1}_{0}n + e^{+} - {}^{1}_{1}H$$

Finally, substituting the known experimental values for W_2 and W_1 (9.5 MeV and 3.8 MeV) and the known values for the masses of the proton and positron, the Joliot-Curies calculated that the mass of the neutron was 1.012 mass units. This value, they said, "is noticeably greater than that of the proton."[198] This was a startling assertion, and although they elaborated on some of its consequences in the conclusion of their paper, they did not explicitly note the direct conflict with Chadwick's result — probably because it was too obvious for words. The discussants also skirted the issue; but, as we shall see, it made a deep impression on Chadwick.

The papers of Dirac, Gamow, and Heisenberg indicate that nuclear electrons and the complex neutron were still very much on their minds, but quite likely the most dramatic moment at the conference occurred when Pauli for the first time proposed his neutrino hypothesis for publication during the discussion immediately following Heisenberg's paper.[199] Pauli's explanation of the continuous beta-ray spectrum on the basis of it, and his blunt assertion that Bohr's idea of nonconservation of energy was "not satisfactory, not even plausible," followed by a list of reasons for his belief, prompted a great deal of further discussion by Chadwick, Meitner, Perrin, and of course Bohr.

We know that this discussion was also a major stimulus for Fermi, who was in the audience, to develop his theory of beta decay following the close of the seventh Solvay Conference.[200] Once completed, his group in Rome persuaded him to send his classic paper to *Nature* as an anti-Hitler gesture, but the editor promptly rejected it for publication, and thus it found its way into the *Zeitschrift für Physik*, where it was received on January 16, 1934.[201] Its opening paragraph leaves no doubt that Fermi saw his theory as reconciling the apparent demand for nuclear electrons with the demand for conservation of energy. As he noted, his theory would assume that only

protons and neutrons are present in the nucleus and, with Heisenberg, that
they represent but two different states of the same particle. The emission of
beta rays, then, occurs "in analogy to the theory of the emission of light
quanta from an excited atom": they would appear when a neutron undergoes
a transition into a proton, electron and neutrino, and disappear in the inverse
process.[202] Segrè recalls that Fermi regarded his theory of beta decay as a
check on his understanding of creation and annihilation operators, and as
quite likely his most important contribution to theoretical physics.[203]

Virtually simultaneously, on January 15, 1934, Curie and Joliot, pursuing
their own researches following the seventh Solvay Conference, astonished
physicists everywhere by reporting their discovery of artificial radioactivity
in the *Comptes rendus*.[204] They found, they said, that when aluminum
($^{27}_{13}$Al) is bombarded with alpha particles, an intermediate step occurs: a
radioactive isotope of phosphorous ($^{30}_{15}$P) and a neutron are formed; then
the isotope decays by positron emission (half-life 3 min 15 sec) into the
isotope of silicon ($^{30}_{14}$Si) identified earlier in Brussels. When Fermi learned of
this far-reaching discovery, he jumped to the brilliant conclusion that not
only alpha particles but also neutrons might produce radioactive isotopes.
He and his group in Rome bombarded element after element in sequence,
and on March 25, 1934, met with success with fluorine.[205]

Meanwhile, another chain of events had been set into motion.[206] Maurice
Goldhaber, leaving Berlin for Cambridge in May 1933 following Hitler's
rise to power, began working with R. H. Fowler in October (just at the time
of the seventh Solvay Conference) and by the end of April 1934 was drafting
a theoretical paper on the role of spin in nuclear reactions. Requiring certain
mass values, he was told that Chadwick was the best source of the information. Goldhaber recalled that during the ensuing discussion he "suddenly
found the courage" to explain to Chadwick how it might be possible to
disintegrate the diplon, the heavy isotope of hydrogen, by bombarding it
with the highly energetic 2.62-million-electron-volt gamma rays emitted by
thorium C″ – a nuclear photoelectric effect. Goldhaber recalled further that
Chadwick seemed to "catch fire" only when it emerged that the mass of
the neutron might be determined accurately by this means – the very topic
that had been so much on Chadwick's mind following the seventh Solvay
Conference.

Approximately six weeks passed before Chadwick one day suddenly
informed Goldhaber that the experiment was working, and invited him to
pursue it further together. Within another two months (August 1934) Chadwick and Goldhaber reported the outcome to *Nature*.[207] The diplon could

indeed be disintegrated by gamma rays into a proton and neutron according
to the reaction

$$_1^2D + h\nu \rightarrow {_1^1}H + {_0^1}n$$

and the mass of the neutron could indeed be determined from the resulting
mass-energy equation

$$m_n = m_D - m_H + W_\gamma - 2W_H$$

where the last term arises because of the virtual identity of the neutron and
proton kinetic energies. They found that the mass of the neutron was 1.008
mass units, definitely greater than the mass of the proton.

These various threads, all bound in one way or another to the seventh
Solvay Conference of October 1933, were woven into the fabric of the
International Conference on Physics of October 1934. Fermi's theory of
beta decay was widely discussed in London and obviously preferred to a
competing theory advanced by Guido Beck.[208] Fermi himself chose to report
on his group's pioneering neutron-bombardment experiments,[209] which had
been directly stimulated by the Joliot-Curies' discovery of artificial radioactivity. It seems clear that these Rome experiments, in addition to their
high intrinsic importance, strongly encouraged physicists to think of the
neutron – now a well-defined projectile – as an elementary particle. A more
subtle influence working in the same direction, indeed ever since Chadwick's
discovery, was the use of a special symbol, n, n^1, $_0n^1$, or $_0^1n$, to designate
the neutron in nuclear reactions.

The Joliot-Curies, who in October 1934 were still mourning the death of
Marie Curie the previous July, were accorded a special welcome in London.
They described their discovery of artificial radioactivity,[210] repeating their
conclusion that in the alpha bombardment of aluminum a radioactive isotope
of phosphorous is formed which subsequently decays by positron emission.
In a separate section of their paper they reported further calculations on the
mass of the neutron, finding 1.0098 and 1.0092 mass units from boron and
aluminum transmutation experiments, and suggesting that 1.010 mass units
be accepted as its value. Once again, this value was significantly greater than
the mass of the proton. "If this is so," they concluded, "the neutron must
be unstable and would be transformed spontaneously" into a proton and
electron with the liberation of 2 million electron volts of energy.[211] Maurice
Goldhaber recalls how startling this conclusion was when it occurred to him
as well.[212]

The theory of the photodisintegration of the diplon, demonstrated by

Chadwick and Goldhaber, formed the subject of H. A. Bethe and R. E. Peierls's paper,[213] and M. L. E. Oliphant cited Chadwick and Goldhaber's value of 1.008 mass units for the mass of the neutron.[214] While differing slightly from the Joliot-Curies' value, Chadwick and Goldhaber's supported a very general conclusion, which must have been evident to everyone in London: far from the possibility of an electron and proton coming together to form a stable combination − the neutron − owing to a mass *defect*, the combination actually suffered from a mass *excess*, was *un*stable, and should decay spontaneously into its constituent electron and proton.

The charges, which were set in Brussels in October 1933 and fired successively in the months that followed, thus landed in London in October 1934 and leveled the nuclear electron hypothesis, even in its guise as the neutron electron hypothesis. K. K. Darrow's contemporary perception of the changing attitude of physicists toward the electron-proton and neutron-proton models of the nucleus was no doubt accurate: "It would be correct to say that the former was the leading scheme until about two years ago; the latter now...."[215]

Very few truly fundamental hypotheses in physics, however, vanish without a trace or are quelled forever. Did not "the successes of Dirac's relativistic theory of the electron" in 1934 suggest that the "dead dream of Lorentz and Poincaré − to give a purely electromagnetic account of the structure of matter − [was] almost at the point of reviving in a new form, an electron-theory account of matter?" asked John A. Wheeler.[216] "The nucleus contains no electrons?" he asked further.

Coming from a family that had fled England in the great religious persecutions of 1620−40, I had grown up in the tradition of silent dissent: silence in public, earnest discussion in private of where one should and where one should not agree with proclaimed doctrine. Therefore, from time to time I took council with Breit, [A. E.] Ruark, and one or two other colleagues about the weaknesses of the standard arguments.[217]

Wheeler's search for deeper and deeper levels of understanding in mid-1930, in common with similar searches by physicists at other times, whether successful or not, excites our admiration and reminds us that the history of physics repeatedly has seen the phoenix rising out of its own ashes.

University of Minnesota

NOTES

I am grateful for comments received from Maurice Goldhaber, Edward M. Purcell, and

Emilio G. Segrè on a draft of this paper. I also thank the National Science Foundation for support received in the early stages of this research.

1 James Chadwick, "Possible Existence of a Neutron," *Nature,* **129** (1932), 312; James Chadwick, "The Existence of a Neutron," *Proc. Roy. Soc.,* A **136** (1932), 692–708; latter reprinted in Robert T. Beyer (ed.), *Foundations of Nuclear Physics* (New York: Dover, 1949), pp. 5–21. See also Norman Feather, "The Experimental Discovery of the Neutron," *Proceedings of the Tenth International Congress of History of Science* [1962], Vol. I (Paris: Hermann, 1964), pp. 135–144; "Chadwick's Neutron," *Contemp. Phys.,* **15** (1974), 565–572.
2 Quoted in W. Pauli, "Zur älteren und neueren Geschichte des Neutrinos," in V. F. Weisskopf (ed.), *Aufsätze und Vorträge über Physik und Erkenntnistheorie* (Braunschweig: Vieweg, 1961), p. 160; reprinted in R. Kronig and V. F. Weisskopf (ed.), *Collected Scientific Papers,* Vol. II (New York: Interscience, 1964), p. 1317 (hereafter cited as *CSP*).
3 Edward M. Purcell, "Nuclear Physics without the Neutron; Clues and Contradictions," *Proc. Tenth Inter. Cong. Hist. Sci.,* Vol. I, pp. 121–132.
4 Werner Heisenberg, "Über den Bau der Atomkerne. I," *Z. Phys.,* **77** (1932), 1–11; "II," *ibid.,* **78** (1932), 156–164; "III," *ibid.,* **80** (1933), 587–596.
5 Joan Bromberg, "The Impact of the Neutron: Bohr and Heisenberg," *Hist. Stud. Phys. Sci.,* **3** (1971), 307–341. Also see, e.g., Hans A. Bethe, Eugene P. Wigner, and Rudolf Peierls in Roger H. Stuewer (ed.), *Nuclear Physics in Restrospect: Proceedings of a Symposium on the 1930s* (Minneapolis: University of Minnesota Press, 1979).
6 Ernest Rutherford, "The Scattering of α and β Particles by Matter and the Structure of the Atom," *Phil. Mag.,* **21** (1911), 669–688; reprinted in J. B. Birks (ed.), *Rutherford at Manchester* (New York: Benjamin, 1963), pp. 182–204, and James Chadwick (ed.), *The Collected Papers of Lord Rutherford of Nelson,* Vol. II (London: Allen and Unwin, 1963), pp. 238–254 (hereafter cited as *CPR*). See also John L. Heilbron, "The Scattering of α and β Particles and Rutherford's Atom," *Arch. Hist. Exact Sci.,* **4** (1968), 247–307.
7 H. Geiger and E. Marsden, "On a Diffuse Reflection of the α-Particles," *Proc. Roy. Soc.,* A **82** (1909), 495–500; reprinted in Birks, *Rutherford,* pp. 175–181.
8 H. Geiger and E. Marsden, "The Laws of Deflexion of α Particles Through Large Angles," *Phil. Mag.,* **25** (1913), 604–623; reprinted in Birks, *Rutherford,* pp. 205–227 (p. 206).
9 Antonius van den Broek, "Die Radioelemente, das periodische System und die Konstitution der Atome," *Phys. Z.,* **14** (1913), 32–41. See also Tetu Hirosige, "The van den Broek Hypothesis," *Jap. Stud. Hist. Sci.,* **10** (1971), 143–162.
10 Niels Bohr, "On the Constitution of Atoms and Molecules. Part II. Systems containing only a Single Nucleus," *Phil. Mag.,* **26** (1913), 476–502; reprinted in L. Rosenfeld (ed.), *On The Constitution of Atoms and Molecules* (Copenhagen: Munksgaard, 1963), pp. 28–54 (p. 29). See also Bohr to Rutherford, June 10, 1913, Bohr Scientific Correspondence (hereafter BSC) in the Archive for History of Quantum Physics (AHQP). There are copies of the AHQP in the American Institute of Physics (AIP) Center for History of Physics, New York: the American Philosophical Society Library, Philadelphia; the Bohr Institute, Copenhagen; the University of California, Berkeley; the University of Minnesota, Minneapolis; the Accademia dei XL, Rome; the Science Museum, London; and the Deutsches Museum, Munich.
11 Geiger and Marsden, "Laws of Deflexion."

[12] A. van den Broek, "Intra-atomic Charge," *Nature,* 92 (1913), 372–373. (His name is spelled here van der Broek, a change made more than once in the literature.)

[13] *Ibid.,* p. 373.

[14] Otto von Baeyer, Otto Hahn, and Lise Meitner, "Uber die β-Strahlen des aktiven Niederschlags des Thoriums," *Phys. Z.,* 12 (1911), 273–279; "Nachweis von β-Strahlen bei Radium D," *ibid.,* 378–379; "Das Magnetische Spektrum der β-Strahlen des Thoriums," *ibid.,* 13 (1912), 264–266.

[15] J. Danysz, "Sur les rayons β de la famille du radium," *Compt. Rend.,* 153 (1911), 339–341; *Radium,* 9 (1912), 1–5.

[16] H. G. J. Moseley, "The Number of β-Particles Emitted in the Transformation of Radium," *Proc. Roy. Soc.,* A 87 (1912), 230–255.

[17] E. Rutherford, "The Origin of β and γ Rays from Radioactive Substances," *Phil. Mag.,* 24 (1912), 453–462; reprinted in *CPR,* Vol. II, pp. 280–287 (pp. 286–287).

[18] See Fajans to Rutherford, Apr. 10, [1913], and O. W. Richardson to Rutherford, June 13, 1913, Rutherford Correspondence (hereafter RC), Cambridge University Library. Van den Broek, "Intra-atomic Charge," p. 373; and Bohr in Rosenfeld, *On The Constitution,* p. 29.

[19] E. Rutherford and H. Robinson, "Heating Effect of Radium and its Emanation," *Phil. Mag.,* 25 (1913), 312–330; reprinted in *CPR,* Vol. II, pp. 312–327.

[20] Frederick Soddy, "Intra-atomic Charge," *Nature,* 92 (1913), 399–400 (on p. 400).

[21] E. Rutherford, "The Structure of the Atom," *Nature,* 92 (1913), 423; reprinted in *CPR,* Vol. II, p. 409.

[22] A. van den Broek, "Intra-atomic Charge and the Structure of the Atom," *Nature,* 92 (1913), 476–478 (p. 477).

[23] E. Rutherford, "The Structure of the Atom," *Phil. Mag.,* 27 (1914), 488–498; reprinted in *CPR,* Vol. II, pp. 423–431.

[24] While Rutherford cited only Soddy in this connection, Soddy's contribution is more tenuous than Fajans'. See Lawrence Badash, *Radioactivity in America: Growth and Decay of a Science* (Baltimore: Johns Hopkins University Press, 1979), esp. Ch. 14, pp. 194–213.

[25] Rutherford, "Structure," *CPR,* Vol. II, p. 431.

[26] J. Chadwick, "Intensitätsverteilung im magnetischen Spektrum der β-Strahlen von Radium B + C," *Ber. Deut. Phys. Ges.,* 12 (1914), 383–391.

[27] E. Rutherford, "The Connexion between the β and γ Ray Spectra," *Phil. Mag.,* 28 (1914), 305–319; reprinted in *CPR,* Vol. II, pp. 473–485.

[28] See E. Rutherford, "Atomic Nuclei and their Transformations," *Proc. Phys. Soc. Lon.,* 39 (1927), 359; reprinted in *CPR,* Vol. III, p. 164.

[29] See the various accounts in the *Dictionary of Scientific Biography* or other standard biographical and autobiographical works.

[30] H. S. Allen, "An Atomic Model with a Magnetic Core," *Phil. Mag.,* 29 (1915), 714–724.

[31] W. D. Harkins, "Recent Work on the Structure of the Atom" (with E. D. Wilson), *J. Amer. Chem. Soc.,* 37 (1915), 1396–1421; "The Nuclei of Atoms and the New Periodic System," *Phys. Rev.,* 15 (1920), 73–94. The latter paper was submitted May 29, 1919, revised Nov. 14, 1919, and published in Feb. 1920. It appears that Harkins had more difficulty in having his work published in physics than in chemistry journals.

[32] A. E. Haas, "Der positive Kern der Atome," *Phys. Z.,* 18 (1917), 400–402.

[33] J. W. Nicholson, "The Radius of the Electron, and the Nuclear Structure of Atoms," *Proc. Phys. Soc. Lon.*, **30** (1918), 1–12. Nicholson's paper was greeted enthusiastically in the discussion by H. S. Allen and Oliver Lodge.

[34] J. J. Thomson, "Problems in Atomic Structure," *Engineering*, **105** (1918), 317–318.

[35] A. W. Steward, "Atomic Structure from the Physico-Chemical Standpoint," *Phil. Mag.*, **36** (1918), 326–336. It was in this paper that Steward introduced the term "isobares" for nuclei of same *Z*, different *A*. While he wished to avoid "isobars" because of its use in meteorology, it was this term that eventually became standard in nuclear physics as well.

[36] Emil Kohlweiler, "Der Atombau auf Grund des Atomzerfalls und seine Beziehung zur chemischen Bindung, zur chemischen Wertigkeit und zum elektrochemischen Charakter der Elemente," *Z. Phys. Chem.*, **93** (1918), 1–42. Emil Kohlweiler, "Konstitution und Konfiguration der Atome," *ibid.*, **94** (1920), 513–541.

[37] H. T. Wolff, "Atomkern und α-Strahlung," *Ann. Phys.*, **60** (1919), 685–701.

[38] W. Kossel, "Über die Zusammensetzung des Atomkerns und seine Neigung zum Zerfall," *Phys. Z.*, **20** (1919), 265–269.

[39] Torahiko Terada, "On a Model of Radioactive Atoms," *Proc. Phys.-Math. Soc. Jap.*, **1** (1919), 185–195.

[40] E. Gehrcke, "Über den Bau der Atomkerne," *Ber. Deut. Phys. Ges.*, **17** (1919), 779–784; E. Gehrcke, "Der Aufbau der Atomkerne," *Sitz. Heidel. Akad. Wiss.* (1920), 1–23.

[41] I. W. D. Hackh, "A Table of the Radioactive Elements which indicates Their Structure," *Phil. Mag.*, **39** (1920), 155–157.

[42] Harkins, "Nuclei" (n. 31), p. 82.

[43] Purcell, "Neutron" (n. 3), p. 125. For a similar evaluation of Harkins's work, see Hans A. Bethe's remarks in Stuewer, *Nuclear Physics* (n. 5), pp. 19–20, 30.

[44] Kohlweiler, "Atombau," p. 11.

[45] E. Rutherford, "Collision of α Particles with Light Atoms. IV. An Anomalous Effect in Nitrogen," *Phil. Mag.*, **37** (1919), 581–587; reprinted in *CPR*, Vol. II, pp. 585–590.

[46] *Ibid.*, *CPR*, p. 589.

[47] See E. Rutherford, "Nuclear Constitution of Atoms," *Proc. Roy. Soc.*, A 97 (1920), 374–400; reprinted in *CPR*, Vol. III, pp. 14–38.

[48] Bohr to Rutherford, July 27, 1920, RC.

[49] Arnold Sommerfeld, *Atombau und Spektrallinien* (Braunschweig: Vieweg, 1919). See "Mathematische Zusätze und Ergänzungen," Sec. 19, "Der Aufbau der Kerne," pp. 535–541.

[50] Rutherford, "Anomalous Effect," *CPR*, Vol. II, p. 589.

[51] E. Rutherford, "The Artificial Disintegration of Light Elements," *Phil. Mag.*, **42** (1921), 809–825; reprinted in *CPR*, Vol. III, pp. 48–62.

[52] P. M. S. Blackett, "The Ejection of Protons from Nitrogen Nuclei, Photographed by the Wilson Method," *Proc. Roy. Soc.*, A 107 (1925), 349–360.

[53] Rutherford, "Nuclear Constitution," *CPR*, Vol. III, p. 17.

[54] E. Rutherford and J. Chadwick, "On the Origin and Nature of the Long-range Particles Observed with Sources of Radium C," *Phil. Mag.*, **48** (1924), 509–526; reprinted in *CPR*, Vol. III, pp. 120–135.

60   ROGER H. STUEWER

55 Rutherford, "Nuclear Constitution," pp. 398, 399; *CPR*, Vol. III, pp. 35, 37.

56 E. Rutherford, "The Building up of Atoms," *Engineering*, 110 (1920), 382.

57 Rutherford, "Nuclear Constitution," *CPR*, Vol. III, p. 34.

58 See Roger H. Stuewer, "William H. Bragg's Corpuscular Theory of X-Rays and γ-Rays," *Brit. J. Hist. Sci.*, 5 (1971), 258–281.

59 See J. L. Glasson, "Attempts to Detect the Presence of Neutrons in a Discharge Tube," *Phil. Mag.*, 42 (1921), 598.

60 *Ibid.*, pp. 596–600; J. Keith Roberts, "The Relation between the Evolution of Heat and the Supply of Energy during the Passage of an Electric Discharge through Hydrogen," *Proc. Roy. Soc.*, A 102 (1922), 72–88.

61 J. Chadwick, "Some Personal Notes in the Search for the Neutron," *Proc. Tenth Inter. Cong. Hist. Sci.* (n. 1), pp. 159–162.

62 Andrade to Rutherford, May 7, 1937, RC.

63 See Goldhaber's comments in Stuewer, *Nuclear Physics*, p. 321.

64 E. Rutherford and J. Chadwick, "Scattering of α-particles by Atomic Nuclei and the Law of Force," *Phil. Mag.*, 50 (1925), 889–913; reprinted in *CPR*, Vol. III, pp. 143–163. Rutherford sketched the potential surfaces in this paper (p. 162), but not the one shown. This more pictorial figure appeared on p. 438 in "Studies of Atomic Nuclei," *Engineering*, 119 (1925), 437–438, based on a Royal Institution Friday lecture of Mar. 27, 1925.

65 P. Debye and W. Hardmeier, "Anomale Zerstreuung von α-Strahlen," *Phys. Z.*, 27 (1926), 196–199.

66 E. Rutherford, "Structure of the Radioactive Atom and Origin of the α-Rays," *Phil. Mag.*, 4 (1927), 580–605; reprinted in *CPR*, Vol. III, pp. 181–202.

67 G. Gamow, "Zur Quantentheorie des Atomkernes," *Z. Phys.*, 51 (1928), 204–212.

68 E. Rutherford, J. Chadwick, and C. D. Ellis, *Radiations from Radioactive Substances* (New York: Macmillan; Cambridge: Cambridge University Press, 1930). See pp. 328 ff. for Gamow's theory, p. 326 for Rutherford's, and p. 327 for Chadwick's comments. We know of Chadwick's authorship of the passage in question, and of Rutherford's response, from Chadwick's interview with Charles Weiner, AIP Center for History of Physics, Apr. 15–21, 1969, transcript p. 49. Emilio Segrè recalls (private communication, Oct. 1979) that Rutherford's satellite model also was found objectionable in Rome, especially by G. Gentile, a close friend of Majorana's and frequent visitor to the Rome group. See G. Gentile, "Sulla teoria dei satelliti di Rutherford," *Atti Accad. Naz. Lincei*, 7 (1928), 346–349. E. N. Gapon, however, in Moscow accepted Rutherford's model still in 1932. See n. 170 below.

69 Key documents in this dispute are J. Chadwick, "Observations concerning the Artificial Disintegration of Elements," *Phil. Mag.*, 2 (1926), 1056–1075, and Chadwick's letters to Rutherford, Dec. 9 and 12 [1927], RC. See also Elizabeth Rona, *How it Came About: Radioactivity, Nuclear Physics, Atomic Energy* (Oak Ridge, Tenn.: Oak Ridge Associated Universities, 1978), pp. 16–20, and Otto R. Frisch, *What Little I Remember* (Cambridge: Cambridge University Press, 1979), p. 64.

70 C. D. Ellis and W. D. Wooster, "The Average Energy of Disintegration of Radium E," *Proc. Roy. Soc.*, A 117 (1927), 109–123.

71 L. Meitner and W. Orthmann, Über eine absolute Bestimmung der Energie der primären β-Strahlen von Radium E," *Z. Phys.*, 60 (1930), 143–155.

72 L. B. Loeb, "Note Concerning the Emission of Beta-rays in Radioactive Change,"

Phys. Rev., **34** (1929), 1212. For one overview of the issues involved see C. S. Wu, "History of Beta Decay," in O. R. Frisch, F. A. Paneth, F. Laves, and P. Rosbaud, (eds.), *Beiträge zur Physik und Chemie des 20. Jahrhunderts* (Braunschweig: Vieweg, 1959), pp. 45–65.
[73] A. F. Kovarik and L. W. McKeehan, *Radioactivity* (*Bull. Nat. Res. Coun.*, Vol. 10, Part 1, No. 51; Mar. 1925), pp. 124–125).
[74] *Ibid.*, p. 124.
[75] E. N. da C. Andrade, *The Structure of the Atom* (New York: Harcourt Brace, 1923), p. 111.
[76] R. A. Millikan, *The Electron* (Chicago: University of Chicago Press, 1917), p. 205; 2nd ed. (1924), p. 207.
[77] Sommerfeld, *Atombau* (n. 49), p. 536.
[78] *Ibid.*, 2nd ed. (1921), p. 569; 3rd ed. (1922), p. 118; 4th ed. (1924), p. 216; 5th ed. (1931), pp. 177 ff.
[79] F. W. Aston, *Isotopes* (London: Arnold, 1922; 2nd ed., 1924). The original term, "isobares" (not isobars), was introduced by A. W. Steward in the paper cited in n. 35 above.
[80] *Ibid.*, pp. 102 and 122, respectively, E. Segrè recalls (private communication, Oct. 1979) that he read this comparison in one of L. Graetz's books and was impressed by it.
[81] N. R. Campbell, *Structure of the Atom* (Cambridge: Cambridge University Press, 1923), pp. 35 ff. Andrade, *Structure*, pp. 110 ff.; 3rd ed. (London: Bell, 1927), pp. 1 ff. Georg von Hevesy and Fritz Paneth, *Lehrbuch der Radioaktivität* (Leipzig: Barth, 1923), pp. 133 ff.; 1st English ed. trans. Robert W. Lawson (Oxford: Oxford University Press, 1926), pp. 151 ff. Jean Perrin, *Atoms*, 2nd English ed. trans. D. L. Hammick (London: Constable, 1923), p. 204, n. 2; pp. 221–225 (Appendix 1921). Hans Pettersson and Gerhard Kirsch, *Atomzertrümmerung* (Leipzig: Akademische Verlagsgesellschaft, 1926), pp. 166 ff. and elsewhere. Hans Pettersson, *Künstliche Verwandlung der Elemente* (Berlin: De Gruyter, 1929), pp. 118 ff. Rutherford, Chadwick, and Eillis, *Radioactive Substances*, esp. pp. 517–525.
[82] W. D. Harkins and W. B. Kay, "An Attempt to Add an Electron to the Nucleus of an Atom," *Phys. Rev.*, **31** (1928), 940–945.
[83] R. Kronig, "Spinning Electrons and the Structure of Spectra," *Nature*, **117** (1926), 550.
[84] See R. Kronig, "The Turning Point," in M. Fierz and V. F. Weisskopf (ed.), *Theoretical Physics in the Twentieth Century: A Memorial Volume to Wolfgang Pauli* (New York: Interscience, 1960), pp. 5–39, esp. pp. 19–28, and B. L. van der Waerden, "Exclusion Principle and Spin," *ibid.*, esp. pp. 209–216. Also see Max Jammer, *The Conceptual Development of Quantum Mechanics* (New York: McGraw-Hill, 1966), pp. 146–152.
[85] G. E. Uhlenbeck and S. Goudsmit, "Ersetzung der Hypothese vom unmechanischen Zwang durch eine Forderung bezüglich des inneren Verhaltens jedes einzelnen Elektrons," *Naturwissenschaften*, **13** (1925), 953–954; "Spinning Electrons and the Structure of Spectra," *Nature*, **117** (1926), 264–265.
[86] Kronig, "Turning Point," p. 28.
[87] Kronig, "Spinning Electrons," p. 550.
[88] W. Pauli, "Zur Frage der theoretischen Deutung der Satelliten einiger Spektrallinien

und ihrer Beeinflussung durch magnetische Felder," *Naturwissenschaften*, **12** (1924), 741–743; reprinted in *CSP*, Vol. II, pp. 198–200.

[89] O. W. Richardson, "Spinning Electrons," *Nature*, **117** (1926), 652.

[90] S. Goudsmit and E. Back, "Feinstrukturen und Termordnung des Wismutspektrums," *Z. Phys.*, **43** (1927), 321–334. For a full review of the literature, see H. Kallmann and H. Schüler, "Hyperfeinstruktur und Atomkern," *Ergeb. Exact. Naturw.*, **11** (1932), 134–175.

[91] R. Kronig, "Der Drehimpuls des Stickstoffkerns," *Naturwissenschaften*, **16** (1928), 335.

[92] For an extensive review, see F. Hund, "Molekelbau," *Ergeb. Exact. Naturw.*, **8** (1929), 147–184.

[93] Kronig, "Drehimpuls," p. 335.

[94] E. Segrè discusses this transition, and Rasetti's work, in "Nuclear Physics in Rome," in Stuewer, *Nuclear Physics*, pp. 35–62, esp. pp. 42–47.

[95] Franco Rasetti, "On the Raman Effect in Diatomic Gases. II," *Proc. Nat. Acad. Sci.*, **15** (1929), 515–519.

[96] See Richardson's remarks in the "Discussion on the Structure of Atomic Nuclei" on Feb. 7, 1929, at the Royal Society, *Proc. Roy. Soc.*, A **123** (1929), 390.

[97] Rasetti, "Raman Effect," p. 519.

[98] Walter Heitler and Gerhard Herzberg, "Gehorchen die Stickstoffkerne der Boseschen Statistik?" *Naturwissenschaften*, **17** (1929), 673–674.

[99] E. P. Wigner and E. E. Witmer, "Über die Struktur der zweiatomigen Molekelspektren nach der Quantenmechanik," *Z. Phys.*, **51** (1928), 859–886.

[100] Heitler and Herzberg, "Stickstoffkerne," p. 673.

[101] *Ibid.*, n. 4.

[102] Wigner's paper was soon published as "Összetett Rendszerek Statisztikája a Quantummechanika szerint," *Magyar Akad. Mat. Term. Ért.*, **46** (1929), 576–582 (German summary, p. 583). It is discussed by Wigner in "The Neutron: The Impact of its Discovery and its Uses," in Stuewer, *Nuclear Physics*, p. 161.

[103] Heitler and Herzberg, "Stickstoffkerne," pp. 673, 674.

[104] Robert S. Mullikan, "Molecular Spectra and Molecular Structure," *Trans. Faraday Soc.*, **25** (1929), 611–645.

[105] *Ibid.*, pp. 644–645. See also Mullikan's extensive article "The Interpretation of Band Spectra. Part IIc. Empirical Band Types," *Rev. Mod. Phys.*, **3** (1931), pp. 89–155, esp. pp. 146–155.

[106] H. Schüler and H. Brück, "Über Hyperfeinstrukturen und Kernmomente," *Z. Phys.*, **58** (1929), 735–741, esp. p. 739n.

[107] Franco Rasetti, "Über die Rotations-Ramanspektren von Stickstoff und Sauerstoff," *Z. Phys.*, **61** (1930), 598–601.

[108] Guido Beck, "Zur Systematik der Isotopen. III," *Z. Phys.*, **61** (1930), 615–618, esp. p. 617.

[109] J. Dorfmann, "Zur Frage über die magnetischen Momente der Atomkerne," *Z. Phys.*, **62** (1930), 90–94 (p. 90). Dorfmann's location is also referred to as the Physical Technical Laboratory. See n. 168.

[110] *Ibid.*, p. 94. Iwanenko and Ambarzumian's report soon appeared. See "Les électrons inobservables et les rayons β," *Compt. Rend.*, **190** (1930), 582–584.

[111] P. Ehrenfest and J. R. Oppenheimer, "Note on the Statistics of Nuclei," *Phys. Rev.*, **37** (1931), 333–338.

[112] P. Güttinger and W. Pauli, "Zur Hyperfeinstruktur von Li$^+$," *Z. Phys.*, **67** (1931), 743–765, esp. p. 754; reprinted in *CSP*, Vol. II, pp. 438–460.

[113] R. M. Langer, "The Absence of Spin in Nuclear Electrons," *Phys. Rev.*, **38** (1931), 837–838. This was also one of the arguments that later impressed John A. Wheeler. See his comments in Stuewer, *Nuclear Physics*, pp. 256–257, 311.

[114] J. H. Van Vleck, *Electric and Magnetic Susceptibilities* (Oxford: Clarendon Press, 1932), pp. 259–60, n. 55.

[115] Cf. Max Jammer, *The Philosophy of Quantum Mechanics* (New York: Wiley, 1974), Ch. 5, pp. 109–158.

[116] E. Rutherford, "Discussion on the Structure of Atomic Nuclei" at the Royal Society, published in *Proc. Roy. Soc.*, A **123** (1929), pp. 373–390 (p. 382).

[117] O. Klein, "Die Reflexion von Elektronen an einem Potentialsprung nach der relativistischen Dynamik von Dirac," *Z. Phys.*, **53** (1929), 157–165.

[118] Cf. Gamow's letters to Bohr, Nov. 25, 1929, and Feb. 25, 1930, BSC, AHQP.

[119] Cf. Bohr to Dirac, Nov. 24, 1929; Dirac to Bohr, Nov. 26, 1929; Bohr to Dirac, Dec. 5, 1929; Dirac to Bohr, Dec. 9, 1929, BSC, AHQP. Also see Bromberg, "Impact" (n. 5), pp. 314–317.

[120] J. Kudar, "Über die Eigenschaften der Kernelektronen," *Phys. Z.*, **32** (1931), 34–37, esp. pp. 36–37. Schrödinger's letters to Bohr of Sep. 25, 1930, and Apr. 29, 1931, BSC, AHQP, e.g., leave no doubt that Schrödinger was having difficulty in guiding Kudar in his work.

[121] F. G. Houtermans, "Neuere Arbeiten über Quantentheorie des Atomkerns," *Ergeb. Exact. Naturw.*, **9** (1930), 123–221, esp. p. 181.

[122] G. Breit, M. A. Tuve, and O. Dahl, "A Laboratory Method of Producing High Potentials," *Phys. Rev.*, **35** (1930), 51–71 (on p. 52).

[123] Bromberg, "Impact," pp. 310–322. For Thomson's papers see "The Disintegration of Radium E from the Point of View of Wave Mechanics," *Nature*, **121** (1928), 615–616; "On the Waves Associated with β-Rays and the Relation between Free Electrons and Their Waves," *Phil. Mag.*, **7** (1929), 405–417.

[124] Bromberg, "Impact," pp. 317–318.

[125] The quotation is taken from Bohr's Faraday Lecture delivered May 8, 1930, and published as "Chemistry and the Quantum Theory of Atomic Constitution," *J. Chem. Soc.*, (1932), 349–384 (p. 380).

[126] Bohr to Mott, Oct. 1, 1929; Rutherford to Bohr, Nov. 19, 1929, BSC, AHQP.

[127] See Bromberg, "Impact," pp. 313–317, for Pauli's and Dirac's responses.

[128] *Ibid.*, p. 325, for quotation.

[129] Kudar, "Eigenschaften," p. 34. Bohr to Kudar, Jan. 28, 1930, BSC, AHQP.

[130] Houtermans, "Neuere Arbeiten," p. 180.

[131] See Laurie M. Brown, "The Idea of the Neutrino," *Phys. Today*, **31** (Sept. 1978), 23–28; C. S. Wu, "The Neutrino," in Fierz and Weisskopf, *Theoretical Physics* (n. 84), pp. 249–303.

[132] Pauli, *Aufsätze* (n. 2), pp. 159–160; reprinted in *CSP*, Vol. II, pp. 1316–1317. Translation in Brown, "Neutrino," p. 27.

[133] E.g., Oppenheimer and J. F. Carlson in 1931, as Brown points out. Pauli, *CSP*,

p. 1319, notes that E. Amaldi told him that Fermi began calling it the "neutrino" in seminars in Rome, evidently after July 1932, as we shall see.

[134] Gamow, "Quantentheorie" (n. 67).

[135] George Gamow, *My World Line: An Informal Autobiography* (New York: Viking, 1970), pp. 59–60.

[136] R. W. Gurney and E. U. Condon, "Quantum Mechanics and Radioactive Disintegration," *Phys. Rev.,* **33** (1929), 127–140.

[137] *Ibid.*, pp. 137–138.

[138] G. Gamow, "Bermerkung zur Quantentheorie des radioaktiven Zerfalls," *Z. Phys.,* **53** (1929), 601–604; written in response to Max von Laue, "Notiz zur Quantentheorie des Atomkerns," *ibid.,* **52** (1928), 726–734.

[139] G. Gamow, "The Quantum Theory of Nuclear Disintegration," *Nature,* **122** (1928), 805–806. J. D. Cockcroft and E. T. S. Walton, "Experiments with High Velocity Positive Ions. II. – The Disintegration of Elements by High Velocity Protons," *Proc. Roy. Soc.,* A **137** (1932), 229–242.

[140] See Gamow's remarks in the "Discussion" (Feb. 7, 1929; n. 116 above), pp. 386–387. Also see "Über die Struktur des Atomkernes," *Phys. Z.,* **30** (1929), 717–720, esp. p. 718; "Mass Defect Curve and Nuclear Constitution," *Proc. Roy. Soc.,* A **126** (1930), 632–644.

[141] Cf. "Über die Struktur," p. 717; the references to the "free nuclear electrons" in "Mass Defect Curve," pp. 632, 637–638 (note also Fig. 2); and the second paragraph of Sec. 1 in "Übergangswahrscheinlichkeiten von angeregten Kernen," *Z. Phys.,* **72** (1931), 492–499.

[142] Cf. "Über die Struktur," p. 719.

[143] Cf. Gamow's 1929–1930 letters to Bohr (n. 118).

[144] George Gamow, *Constitution of Atomic Nuclei and Radioactivity* (Oxford: Clarendon Press, 1931).

[145] See interview with Charles Weiner, AIP Center for History of Physics, Apr. 25, 1968, transcript pp. 33 ff.

[146] Gamow, *Atomic Nuclei,* p. 1.

[147] *Ibid.*, pp. 2–5, 27, 29, 56, 69.

[148] *Ibid.*, p. 5.

[149] Léon Rosenfeld, "Nuclear Physics, Past and Future," in M. Nève de Mévergnies, P. Van Assche, and J. Vervier (eds.), *Nuclear Structure Study With Neutrons* (Amsterdam: North-Holland, 1966), p. 484.

[150] Cf. Georges Fournier, "Sur une classification nucléaire des atomes en relation avec leur genèse possible et leur désintégration radioactive," *J. Phys. Rad.,* **1** (1930), 194–205 (see, e.g., p. 198); R. M. Langer and N. Rosen, "The Neutron," *Phys. Rev.,* **37** (1931), 1579–1582; Harold C. Urey, "The Natural System of Atomic Nuclei," *J. Amer. Chem. Soc.,* **53** (1931), 2872–2880 (see, e.g., p. 2876); W. D. Harkins, "The Periodic System of Atomic Nuclei and the Principle of Regularity and Continuity of Series," *Phys. Rev.,* **38** (1931), 1270–1288; and Samuel D. Bryden, Jr., "The Structure of the Nucleus and its Total Moment of Momentum," *ibid.,* 1989–1994 (no specific mention of term "neutron" but use of concept of "equivalent protons" in building up nuclei). Harkins, of course, on p. 1288 of his paper, was careful to assert his priority over Langer and Rosen. Richard Swinne, "Neutron, das nullte Element," *Z. tech. Phys.,* **13** (1932), 279, claimed that W. Nernst had coined the term "neutron" in 1903.

151 See, e.g., the letter of J. D. Cockcroft to E. T. S. Walton and P. I. Dee, June 24, 1933, RC, in which Cockcroft reports that during his visit to Chicago Harkins explained how he had discovered the neutron already in 1915.
152 Irène Curie and F. Joliot, "Émission de protons de grande vitesse par les substances hydrogenées sous l'influence des rayons γ tres pénétrants," *Compt. Rend.* **194** (1932), 273–275. See also. W. Bothe and R. Becker, "Künstliche Erregung von Kern-γ-Strahlen," *Z. Phys.*, **66** (1930), 289–306.
153 Quoted in Feather, "Chadwick's Neutron" (n. 1), p. 569.
154 See Chadwick's letter to Bohr, Feb. 24, 1932, BSC, AHQP.
155 Chadwick, "Possible Existence" (n. 1).
156 See the entry for meeting 302 of the Kaptiza Club, Feb. 23, 1932, AHQP.
157 See n. 154.
158 "Discussion on the Structure of Atomic Nuclei," *Proc. Roy. Soc.*, A **136** (1932), 735–762; Chadwick's remarks are on pp. 744–748.
159 Chadwick, "Existence" (n. 1).
160 *Ibid.*, p. 700.
161 *Ibid.*, p. 702.
162 *Ibid.*, p. 704.
163 *Ibid.*, p. 705. Chadwick cites unpublished discussions with Bohr in Copenhagen, and two published reports by H. S. W. Massey of 1932.
164 *Ibid.*, p. 706.
165 "Discussion" (n. 158).
166 *Proc. Roy. Soc.*, A **142** (1933), 1–25.
167 *Ibid.*, p. 15.
168 D. Iwanenko, "The Neutron Hypothesis," *Nature*, **129** (1932), 798.
169 Dorfmann, "Frage" (n. 109).
170 D. Iwanenko, "Sur la constitution des noyaux atomiques," *Compt. Rend.*, **195** (1932), 439–441. See also E. Gapon and D. Iwanenko, "Zur Bestimmung der Isotopenzahl," *Naturwissenschaften*, **20** (1932), 792–793, and E. N. Gapon," Zur Theorie des Atomkernes," *Z. Phys.*, **79** (1932), 676–681; "II," *ibid.*, **81** (1933), 419–424; "III," *ibid.*, **82** (1933), 404–407. In the first of these three papers, Gapon, in Moscow, accepts the neutron as an electron-proton compound, following Rutherford's conception of neutral satellites surrounding the central nuclear core.
171 Kallmann and Schüler, "Hyperfeinstruktur" (n. 90), p. 154.
172 *Ibid.*, p. 158.
173 F. Perrin, "L'existence des neutron et la constitution des noyaux atomiques légers," *Compt. Rend.*, **194** (1932), 1343–1346. Following Iwanenko's paper and others, Perrin extended his ideas in July 1932 to consider some implications for spin if nuclei were composed of alpha particles, "demi-helions" (2 protons + 1 electron), and neutrons (1 proton + 1 electron). See "La constitution des noyaux atomiques et leur spin," *ibid.*, **195** (1932), 236–237.
174 G. Fournier, "Sur la composition des noyaux atomiques," *Compt. Rend.*, **194** (1932), 1482–1483. Marie Curie, "Les Rayons des corps radioactifs en relation avec la structure nucléaire," in Robert de Valbreuze, ed., *Comptes Rendus du Congrès International d'Électricité, Première Section* (Paris: Gauthier-Villars, 1932), pp. 809–832, esp. p. 830, where she refers to Chadwick's neutron as "formed by a particularly intimate association of a proton and electron."

[175] Bromberg, "Impact" (n. 5), p. 331.

[176] E.g., H. A. Bethe, E. P. Wigner, and R. Peierls in Stuewer, *Nuclear Physics*. Heisenberg, "I," "II," and "III" (n. 4).

[177] D. M. Brink, *Nuclear Forces* (Oxford: Pergamon, 1965), p. 154, n. iii, points out that Heisenberg's belief that the internal structure of the neutron influences the scattering of gamma rays by nuclei was proved to be mistaken by experiments of Meitner and Hupfeld.

[178] Heisenberg, "I," esp. pp. 1–3, 8.

[179] Heisenberg, "II," p. 163.

[180] *Ibid.*, p. 164.

[181] Heisenberg, "III," pp. 594–596.

[182] W. D. Harkins, "The Neutron, Atom Building and Nuclear Exclusion Principle," *Proc. Nat. Acad. Sci.*, 19 (1933), 307–318, esp. p. 312, where he states that the possibility that the "Harkins-Masson nuclear formula" "might be made the basis of a theory of the nucleus was pointed out by the writer to Heisenberg in 1929, and he has used it as the basis of his recent nuclear theory." On p. 313 Harkins claims that another of his relations, of 1921, "has just been recognized by Heisenberg, who uses it as the most important in his theory."

[183] Quoted in Stuewer, *Nuclear Physics*, p. 48.

[184] *Ibid.*, pp. 43–47, where Segrè notes their importance and influence.

[185] *Ibid.*, p. 48. The Proceedings of the Paris conference are cited in n. 174 above. Segrè has recently confirmed (private communication, Oct. 4, 1979) that his recollection and impression is that the nuclear electron hypothesis was really dead in Rome by late spring 1932.

[186] Enrico Fermi, "État actuel de la physique du noyau atomique," in *Comptes Rendus du Congrès* (n. 174), pp. 789–807. See also n. 133 above.

[187] Ettore Majorana, "Über die Kerntheorie," *Z. Phys.*, 82 (1933), 137–145.

[188] Hans Bethe, "The Happy Thirties," in Stuewer, *Nuclear Physics*, p. 13.

[189] Rudolf Peierls, "The Development of Our Ideas on the Nuclear Forces," in Stuewer, *Nuclear Physics*, pp. 183–211.

[190] Merle Tuve, "The Atomic Nucleus and High Voltages," *J. Franklin Inst.*, 216 (1933), 1–38.

[191] K. T. Bainbridge, "The Masses of Atoms and the Structure of Atomic Nuclei," *J. Franklin Inst.*, 215 (1933), 509–534 (p. 525). K. K. Darrow, "Contemporary Advances in Physics, XXVI. The Nucleus, First Part," *Bell Sys. Tech. J.*, 12 (1933), 288–330; offprint pp. 1–43 (p. 4).

[192] *Structure et Propriétés des Noyaux Atomiques, Rapports et Discussions du Septième Conseil de Physique* (Paris: Gauthier-Villars, 1934). *International Conference on Physics, London 1934. Papers and Discussions.* Vol. I. *Nuclear Physics* (London/Cambridge: The Physical Society and Cambridge University Press, 1935).

[193] J. Chadwick, "Diffusion anomale des particles α. Transmutation des éléments par des particles α. Le neutron," *Structure et Propriétés*, pp. 81–112; "The Neutron" occupies pp. 98–112.

[194] *Ibid.*, p. 102.

[195] *Ibid.*, pp. 103–110.

[196] I. Curie and F. Joliot, "Rayonnement pénétrant des atoms sous l'action des rayons α," *Structure et Propriétés*, pp. 121–156.

[197] *Ibid.*, pp. 155–156.
[198] *Ibid.*, p. 156.
[199] *Ibid.*, pp. 324–325.
[200] See E. Segrè's comments in Stuewer, *Nuclear Physics*, p. 49.
[201] "Versuch einer Theorie der β-Strahlen. I," *Z. Phys.*, 88 (1934), 161–177; reprinted in Beyer, *Foundations* (n. 1), pp. 45–61.
[202] *Ibid.*, pp. 161–162.
[203] See Segrè's comments in Stuewer, *Nuclear Physics*, p. 49.
[204] "Un nouveau type de radioactivité," *Compt. Rend.* 198 (1933), 254–256; reprinted in Beyer, *Foundations*, pp. 39–41.
[205] E. Fermi, E. Amaldi, B. Pontecorvo, F. Rasetti, and E. Segrè, "Azione di sostanze idrogenate sulla radioattività provocata da neutroni," *Ricer. Scient.*, 5 (1934), 282–283. See also Segrè's comments in Stuewer, *Nuclear Physics*, pp. 50–52.
[206] The succeeding account follows that given by Maurice Goldhaber, "The Nuclear Photoelectric Effect and Remarks on Higher Multipole Transitions: A Personal History," in Stuewer, *Nuclear Physics*, pp. 83–106, esp. pp. 85–89.
[207] J. Chadwick and M. Goldhaber, "A 'Nuclear Photo-effect': Disintegration of the Diplon by γ-Rays," *Nature*, 134 (1934), 237–238. Modern values for the mass of the neutron m_n and the mass of the proton m_p, based on the 16_O scale are as follows: m_n = 1.008982 amu and m_p = 1.007593 amu. See W. E. Burcham, *Nuclear Physics: An Introduction* (New York: McGraw-Hill, 1963), pp. 728–729.
[208] G. Beck, "Report on Theoretical Considerations Concerning Radioactive β-Decay," in *International Conference* (n. 192), pp. 31–42. C. D. Ellis, in his paper, noted both Beck's and Fermi's theories and stated simply: "In discussing the β-ray disintegrations we shall use Fermi's theory." *Ibid.*, p. 47.
[209] E. Fermi, "Artificial Radioactivity Produced by Neutron Bombardment," *ibid.*, pp. 75–77.
[210] I. Curie and F. Joliot, "Artificially Produced Radio-Elements," *ibid.*, pp. 78–86.
[211] *Ibid.*, p. 83.
[212] See his comment in Stuewer, *Nuclear Physics*, p. 88.
[213] H. A. Bethe and R. E. Peierls, "Photoelectric Disintegration of the Diplon," in *International Conference*, pp. 93–94.
[214] M. L. E. Oliphant, "Transformation Effects Produced in Lithium, Heavy Hydrogen and Beryllium, by Bombardment with Hydrogen Ions," *Ibid.*, p. 148.
[215] K. K. Darrow, "Contemporary Advances in Physics, XXIX. The Nucleus, Fourth Part," *Bell Sys. Tech. J.*, 14 (1935), 285–321; separate offprint pp. 1–37 (p. 32).
[216] These questions occur in John Wheeler's provocative section "What is the Role of the Electron in Nuclear Physics?" in his paper "Some Men and Moments in the History of Nuclear Physics: The Interplay of Colleagues and Motivations," in Stuewer, *Nuclear Physics*, pp. 217–306 (pp. 254–255).
[217] *Ibid.*, pp. 255–256.

ROBERT KARGON

THE EVOLUTION OF MATTER: NUCLEAR PHYSICS, COSMIC RAYS, AND ROBERT MILLIKAN'S RESEARCH PROGRAM

When Robert A. Millikan was discharged from the Army after World War I he faced a difficult, but sweet, career problem. Before the war, he had completed a series of investigations that many believed at the time would bring him the Nobel Prize. He had just been demoblized from a post that brought him into contact with the leading scientists and scientific entrepreneurs of America. Few prior research commitments weighed upon him. He returned to Chicago and was wooed by the astrophysicist George Ellery Hale to become president of a vigorous educational enterprise in Pasadena, California, soon to be named the California Institute of Technology.[1] It became obvious that his new series of investigations would have to be on signigicant problems, for they would have to mirror his newly elevated status in the profession.

His previous successful researches had aimed at the heart of new theories of matter and radiation: the nature of the electron and the character of light. He wished now to turn to the pressing problem of probing inside the atom in order to determine, precisely, the map of the atom's interior and if possible to gain some further insight into what had always pricked his imagination, the evolution and transformation of the elements.

The decade prior to Millikan's arrival in Pasadena had been one of the most exciting and fruitful in the history of physics. The plethora of data and speculation opened by the exploitation of the spectroscope, by the discovery of X rays, and by investigations into radioactivity was beginning to come under some reasonable control. Rutherford's nuclear atom and the physics of the quantum were joined in Niels Bohr's landmark atomic model. Bohr's atom was a spectroscopic atom: it was designed to cope with the enormous quantity of line-spectra data which physicists had long intuited would yield information about atomic structure.[2] Millikan later remarked that when "it was devised, spectroscopy was a veritable dark continent in physics. With the aid of the Bohr atom the dark continent in physics has become the best explored and best understood, the most civilized portion of the world of physics. It has been an exciting game of exploration."[3]

The "exciting game" was further stimulated by the discoveries of Henry G. J. Moseley, the British physicist who, working along lines first laid out

69

William R. Shea (ed.), Otto Hahn and the Rise of Nuclear Physics, 69–89.
Copyright © 1983 by D. Reidel Publishing Company.

by the X-ray crystallographers Max von Laue and the Braggs, accurately
determined the frequencies of the X rays characteristic of the elements.
Moseley showed that the square roots of these frequencies are related in a
simple way. They constitute a simple aritmetic progression: each member
of the series is obtained from the previous member by adding the same
quantity. Morsley saw at once that his formulas could be used to test the
periodic table, for his series of X-ray frequencies (with a few exceptions)
follows the chemists' series of atomic weights.[4] Furthermore, radioactivity
research produced evidence which demonstrated that when a substance loses
a doubly charged positive particle (an alpha) it moves two places to the left
in this periodic table, and when it loses a single negative particle it moves a
place to the right. The chemical nature of a substance depends, therefore,
upon its atomic number, or the number of positive charges in its nucleus.
The periodic table, on the basis of Moseley's work, could now be recons-
tructed alone physical lines, utilizing this notion of atomic number. "There
is," Moseley stated, "in the atom a fundamental quantity which. . . increases
by regular steps. . . from one element to the next. This quantity can only be
the charge on the central positive nucleus."[5]

It was Millikan's intention to follow along the path traveled by Moseley. He
signaled his intentions in his presidential address before the Physical Society
in December 1916. The address, "Radiation and Atomic Structure" sketches
his strategy. "It has been chiefly," Millikan asserted, "the facts of radiation
which have provided reliable information about the inner structure of the
atom itself." Charles Crover Barkla, Rutherford, and, above all, Moseley,
have furnished evidence concerning the "electronic" and nuclear constituents
of the atom: "In a research which is destined to rank as one of the dozen
most brilliant in conception, skillful in execution, and illuminating in results
in the history of science, a young man but twenty-six years old [Moseley]
threw open the windows through which we can now glimpse the sub-atomic
world with a definiteness and certainty never dreamed of before."[6] The task
before the physics community is articulation of the atomic model of Niels
Bohr:

If then the test of truth in a physical theory is large success both in the prediction of
new relationships and in correctly and exactly accounting for old ones, the theory of
non-radiating orbits is one of the well-established truths of modern physics. For the
present at least it is truth and no other theory of atomic structure can be considered
until it has shown itself able to approach it in fertility. I know of no competitor which
is as yet in sight.[7]

However, the Bohr theory, unmodified, applies to hydrogen and helium; for heavier elements, "the radiations give us no information about conditions or behaviors of the external electrons which have to do with the phenomena of valency."[8]

The new program would deal directly with the questions of atomic structure, investigated through spectroscopic studies, and would later be joined by studies of the "evolution" of the elements, examined through investigations of "penetrating radiation," or, as this was later known, "cosmic rays." His plan was described in the joint grant proposal to the Carnegie Corporation in 1921, wherein Millikan (along with Hale and Arthur Noyes) wrote:

Matter occurs in nature under the widest variety of composition and form. The physicist, who approaches the complex problem by the simplest and most direct route, deals only with the chemical elements, and evolves powerful methods of research which enable him to penetrate to the core of the atom, to visualize the electrons swinging in their orbits, and to remove them one by one for detailed study.[9]

The first major step in the new program was the development of a new technique. As early as 1905 Millikan had been working on problems related to sparking potentials in vacua,[10] and had found that the spark discharge emitting ultraviolet light of a condenser could be maintained under high potentials in such vacua. According to Millikan, it "occurred" to him that such discharges, labelled "hot spark discharges," could solve a troublesome spectroscopic problem. One difficulty in mapping the spectra of extreme ultraviolet light emitted from atoms was the high absorption of very short wavelength radiations by the fluorite window and prism spectroscopic equipment then in general use.[11] Millikan believed he could circumvent these difficulties using a vacuum spectrometer employing high vacua, hot sparks, and a concave reflecting grating. In a series of papers with R. A. Sawyer and Ira Bowen, Millikan was able to make a considerable extension of the map of the ultraviolet spectrum; that is, he was able to photograph, measure the wavelength, and analyze the atoms of light elements and multiply ionized atoms of the heavier atoms. They found about a thousand new lines and showed that their wavelengths were predicted by the Bohr theory.[12] Millikan and Bowen established the essential unity of the optical and X-ray spectra, demonstrating, in Millikan's words, that "optical spectra are quite like X-ray spectra in that large gaps occur between frequencies due to electrons in successive rings or shells.[13]

Millikan also worked with his graduate students on the emission of electrons from cold metal surfaces from points at which the field gradient is large. Although Millikan and B. E. Shackelford concluded in 1920 that their "experiments indicate that the discharge is conditioned by surface impurities and cast doubt upon the conclusion that there is a particular field strength at which electrons begin to be pulled out of a pure metal,"[14] Carl Eyring, then a Caltech graduate student, and Millikan found in 1926 that the *field current* (a term they apparently introduced) depended upon the local field gradient at the point of emission and not on the total potential difference along the wire. Furthermore, its behavior was incompatible with the assumption that the field current was a thermionic current in a strong electrical field.[15] The quantum-mechanical explanation for the phenomenon was given by J. Robert Oppenheimer and by R. H. Fowler and L. Nordheim, who independently explained field current as leakage through a potential barrier.[16]

The third segment of the program dealt with what was usually termed at the time the "penetrating radiation" – what was to become known by Millikan's appellation "cosmic rays." At the beginning of the century several researchers found that normal air in the laboratory was slightly ionized.[17] Shortly thereafter, Ernest Rutherford and H. Cooke and John McLennan and Eli Burton, all working in Canada, demonstrated a marked reduction in ionization when the detecting vessel is shielded by lead or some other absorbing material, indicating that the ionization arises from some outside radiation which was able to penetrate thin metal walls.[18] It was supposed for a long time afterwards that the ionization was caused by radiation from small amounts of radioactive material in rocks and soil. In 1909 K. Kurz reviewed all the evidence and concluded that nothing contradicted the hypothesis that the penetrating radiation was caused by such radioactive impurities in the earth's crust.[19]

A corollary of this view – that the radiation ought to decrease at high altitudes – was tested atop high towers and by daring balloon ascents by K. Bergwitz and A. Gockel, accounts of which were published in 1910 and 1911.[20] The results were uncertain, but Gockel uncovered a strange anomaly: at 4500 meters the intensity of the ionization was actually *greater* than at the earth's surface. Victor Hess and (independently) Werner Kolhörster mounted numerous balloon ascents and improved the apparatus.[21] They were able to show that above 2000 meters the penetrating radiation increases and above 3000 meters there is a marked rise in intensity. Hess and Kolhörster were able to attain heights of 5200 and 9000 meters respectively. Hess suggested that there is a very penetrating radiation which enters the atmosphere from above,[22] the first glimmer of what were to be known as cosmic rays. But the nature of this radiation, its absorption properties, and even its existence were matters of great uncertainty in the decade following Hess's suggestion.

Millikan became interested in the penetrating radiation shortly before America's involvement in the World War. His concern is not surprising: he had long been interested in radioactivity, and the research programs concerning the penetrating radiation had been conceded to be a part of the study of radioactive phenomena. Hess and his co-worker M. Kofler insisted, however, that the rays came from outside the atmosphere and still believed the rays to be radium and thorium emanations.[23] When Millikan was demobilized from the Army, he resolved to return to the questions raised by Hess and the others. In a letter to Robert Woodward of the Carnegie Institution of Washington dated May 6, 1919, Millikan requested assistance for a project of research "upon the penetrating radiation and the degree of ionization existing in the regions of the atmosphere between elevations of five miles and twenty miles." The investigations were to be carried out "by designing and sending up in sounding balloons self-recording instruments of a new type." By means of innovative apparatus — the use of unmanned flight and self-recording instrumentation — Millikan would have been able, once again, to bring clarity and precision into a scientific problem area. "We have already indications of a penetrating radiation which has its source outside the earth, he wrote." If such a source does exist it is a matter of extraordinary importance to have it revealed and studied. I know of no way other than that suggested in which definite knowledge as to the existence or non-existence of such a source can be obtained.[24] It seems clear that Millikan saw his problem first to demonstrate what had merely been indicated, the existence of the extra-atmospheric penetrating radiation, and second, if the radiation did exist, to determine its characteristics. The Carnegie Corporation grant proposal submitted in 1921 by Hale, Millikan, and Noyes confirms this view. In the section titled "The Research Program from the Standpoint of Physics," very little space is devoted to the penetrating radiation; and the proposal merely states that it is Millikan's intention to "study the penetrating radiations of the upper air to determine whether these are of cosmic or of terrestrial origin."[25]

What the Carnegie proposal emphasizes, however, is another, as yet unlinked, interest of Millikan's. From his early research during the first decade of the century, Millikan's imagination had been captured by radioactivity, the transformation of the elements, and its potential insights into the construction of matter. He was of course not alone; many physicists shared these concerns, and the brilliant work of Moseley reinforced their concerns.

Moseley's work revived interest in Prout's hypothesis, which had been advanced a century earlier. The chemist William Prout had argued that atomic weights are multiples of that of hydrogen, and as a consequence hydrogen may be taken to be the primordial element. Prout's speculation foundered upon the rock of precise measurement: the multiple relationships are not exact. But twentieth-century physicists were confident of resolving all difficulties, and the assumption was widely held that the elements were built up out of hydrogen nuclei (protons) and electrons. Arnold Sommerfeld's magisterial *Atombau und Spektrallinien* endorsed Prout in this way: "The atoms of the various elements must be similarly constructed out of identical units."[26] Millikan himself put it a bit more romantically in his book *The Electron* of 1917: "It looks as if the dream of Thales of Miletus has actually come true and that we have found not only a primordial element out of which all substances are made, but that primordial element is hydrogen itself."[27]

The exact work of F. W. Aston with the mass spectograph served only to reinforce interest in atom-building. Aston's "whole number rule" — that with the exception of hydrogen, atomic weights of the elements are whole numbers — "removes the only serious objection to a unitary theory of matter."[28] Hydrogen was measured at very slightly more than unity; the "lost" mass in the building up of heavier elements from the proton was explained away by a "packing effect" which was originally viewed as an electromagnetic contraction.[29] Whatever the mechanism of packing, it was widely believed that in the construction of the heavier elements some mass remained to be accounted for. Aston's influential book *Isotopes* of 1922 was merely recording the commonly held belief that "we may consider it absolutely certain that if hydrogen is transformed into helium a certain quantity of mass must be annihilated in the process. . . . Should the research worker of the future discover some means of releasing this energy in a form which could be employed, the human race will have at its command powers beyond the dreams of science fiction."[30]

This revived concern with the construction of the elements from hydrogen nuclei and electrons was seen not only as possible in principle but as a process actually occurring in the heavens and possibly capable of realization on earth. Ernest Rutherford, invited by Hale to appear before the National Academy of Sciences in 1915, spoke on "The Constitution of Matter and the Evolution of the Elements" and splendidly summed up the views of many physicists and astrophysicists:

It has been long thought probable that the elements are all built up of some fundamental substances, and Prout's well-known hypothesis that all atoms are composed of hydrogen is one of the best-known examples of this idea. . . . In the hottest stars the spectra of hydrogen and helium predominate, but with decreasing temperature the spectra become more complicated and the lines of the heavier elements appear . . . . it is supposed that the light elements combine with decreasing temperature to form the heavier elements. There is no doubt that it will prove a difficult task to bring about the transmutation of matter under ordinary terrestrial conditions. The enormous evolution of energy which accompanies the transformation of radioactive matter affords some indication of the great intensity of the forces that will be required to build up lighter into heavier atoms.[31]

Millikan had long been deeply interested in radioactivity and the problem of the evolution of the elements. His first research program in the new physics concerned these subjects. In a review of the field for *Popular Science Monthly* in 1904 he plainly revealed his enthusiasm: "the dreams of the ancient alchemists are true, for the radio-active elements all appear to be slowly but spontaneously transmuting themselves into other elements.[32] Millikan's noteworthy conclusion lays down a theme to which he would return many times in the succeeding two decades:

. . . the studies of the last eight years upon radiation seem to indicate that in the atomic world also, at least *some* of the heaviest and most complex atomic structures are tending to disintegrate into simpler atoms. The analogy suggests the profoundly interesting question, as to whether or not there is any natural process which does, among the atoms, what the life process does among the molecules, i.e., which takes the simpler forms and builds them up into more complex ones.[33]

The "profoundly interesting question" of discovering a natural process of building complex nuclei from simpler ones was part of Millikan's first foray into artificial transmutation (by his own testimony) in 1912. Millikan and his student G. Winchester believed that they had produced hydrogen ions from aluminum by high-voltage discharges.[34] After Sir William Ramsay, Professor Norman Collie, and H. Patterson announced that they thought they had produced helium and neon using electrical discharges,[35] Winchester, though not Millikan, ventured forth into print, and suggested that while helium and neon were merely occluded gases, *hydrogen* can be liberated from aluminum "in somewhat the same manner as the α particle is disintegrated from radium.[36]

The revolutionary work of Rutherford and James Chadwick in 1919–1921 on artificial nuclear transformations created fascinating new possibilities.[37] Deeper understanding of the atom-building process awaited further penetration into the mysteries of the nucleus. Studies of artificial transmutations, it now seemed, offered clues to these mysteries. Millikan followed Rutherford's work with his usual keen and close interest.[38] The possibilities which he foresaw were hinted at in an address given in Washington, shortly before his decision to come to Caltech on a permanent basis. His exuberance about the new world of physics on the verge of being opened was scarcely veiled:

... we have been forced to admit for the first time in history not only the possibility but the fact of the growth and decay of the elements of matter. With radium and with uranium we do not see anything but the decay. And yet somewhere, somehow, it is almost certain that these elements must be continually forming. They are probably being put together now in the laboratories of the stars. . . . Can we ever learn to control the process? Why not? Only research can tell. What is it worth to try it? A million dollars? A hundred million? A billion? It would be worth that much if it failed, for you could count on more than that amount in by-products. And if it succeeded, a new world for man![39]

It was this "new world for man" and the attempt to come to grips with it which contributed to his interest in the construction of the High Voltage Laboratory and to which Hale referred when he wrote of Millikan's interest in "This possibility which no other laboratory could match," that is, a laboratory which would enable Millikan "to bust up some of his atoms."[40] That Millikan intended to use the high-voltage capacities of the proposed laboratory to probe the nucleus as well as for other problems is borne out by the prominence afforded these intentions in the major grant proposals constructed by him and by Hale for the Carnegie Corporation of New York in mid-1921. In a preliminary proposal submitted to the Corporation's president, James Angell, on June 4, 1921, Hale referred to the High Voltage Laboratory more explicitly as "especially adapted for Dr. Millikan's researches on the breaking-down of atoms and the resolution of the chemical elements into simpler components."[41]

In the fuller grant proposal submitted to the Corporation, Hale revealed that nuclear transformations were to occupy a major place among the concerns of the California Institute of Technology, now emerging as an ambitious center for physical-science research. The heart of the proposal was a joint attack on problems of radiation and matter by the Gates Laboratory and by Hale's Mount Wilson Observatory as well as by Millikan's soon-to-be opened Norman Bridge Laboratory.[42] In their brief summary, Hale, Noyes, and Millikan pointed to several exciting new areas for exploration: radioactive transmutations and the existence of a basic element (hydrogen), direct attention to the possibility of artificial transmutation in the laboratory and the probability that "in the stars the heavier elements are being built up from the lighter ones, a process not yet realized on earth." The projected physics program included high-potential work bearing "upon the nature of atoms and their possible transformations into one another."[43] It is clear that program was to encompass both heaven and earth, through a laboratory attack involving Millikan's plan to "bust up" atoms and also through investigations of stellar processes by the staff of the Mount Wilson Observatory.

A press release, which apparently never saw the light of day, was prepared by Sam Small, Jr., based upon interviews with Millikan in 1922. It reported: "It is the present task of the scientists at the Institute of Technology, by means of fundamental research working through several directions, to smash the nucleus of the atom and find out what it is made of."[44] Millikan's optimism along these lines continued into 1923. In an article published that year in *Scribner's Magazine*, "Gulliver's Travels in Science," Millikan reported upon the present state of knowledge of transmutable elements and asked rhetorically: "Does the process go on in both directions, heavier atoms being formed as well as continually disintegrating into lighter ones? Not on earth as far as we can see. Perhaps in God's laboratories, the stars. The key question, however, remained:

Can we on earth artificially control the process? To a very slight degree we know already how to *disintegrate* artificially, but not as yet how to build up. As early as 1912, in the Pyerson Laboratory at Chicago, Doctor Winchester and I thought we had good evidence that we were knocking hydrogen out of aluminum and other metals by very powerful electrical discharges in vacuo. We still think our evidence to be good.[45]

Millikan has here made an extraordinary claim. He believed, as of 1923, that he and Winchester had anticipated Rutherford's artificial transmutation, not as Rutherford had done, with alpha-particle bullets, but with high-voltage discharges.[46] No wonder then that Millikan was eager to erect and to utilize a high-voltage laboratory with its million-volt transformer. "How much farther," he continued, "can we go into this artificial transmutation of the elements? This is one of the supremely interesting problems of modern physics *upon which we are all assiduously working*."[47]

The million-volt transformer was built for the High Voltage Laboratory by Royal Sorensen, professor of electrical engineering at Caltech. Four 250,000-volt 50-cycle Westinghouse transformers were installed, cascade fashion, on steel frames, supported by porcelain insulators.[48] But whether the cascade transformer itself was utilized by Millikan for the study of nuclear transformations is doubtful. He more likely made trials similar to those published by Winchester in 1914. Winchester had built vacuum tubes with an aluminum cathode and platinum anode, and placed high potentials (at that time about 100,000 volts) across them. He generated helium, neon,

and hydrogen gases. After careful study he concluded that the neon and
helium had been occluded at or near the surface of the electrodes, but that
the hydrogen was possibly to be understood as a disintegration product of
aluminum. It is reasonable to assume that given his stated interest, Millikan
may have made further attempts along these lines in 1922 and 1923, but
concluded that the hydrogen, previously identified spectroscopically, was not
in fact a disintegration product.[49]

Millikan's plans to smash the nucleus with his high voltages were therefore
never successfully prosecuted. In retrospect, with all the advantages of hindsight accrued, even the discussion may seem bizarre. But as we have seen, in
the context of the very fluid situation of the early 1920s, Millikan was by no
means unjustified in risking some time and research capital. Transmutation
was a popular enterprise in the 1920s. Successful attempts at turning base
metals into gold using medium and high voltages were reported by A. Miethe
and H. Stammreich and H. Nagaoka.[50] Lead was reportedly transmuted
into mercury by A. Smits of Amsterdam.[51] These claims were vehemently
opposed by F. Haber and by F. Aston.[52] Millikan was perhaps too wise an
experimentalist to move very far out on this limb.

Instead he turned his energies to a more promising subject that had in-
trigued him for the better part of a decade. Millikan's search for fuller understanding of the atomic nucleus and its transformations helped to concentrate
his efforts in the area which came to be known by the name he himself gave
it—cosmic rays.

That a fuller understanding of atom-building and the study of astrophysical phenomena were widely seen during the early 1920s to be closely
related is reflected in Rutherford's presidential address before the Liverpool
meeting of the British Association for the Advancement of Science in September 1923. In this address Rutherford discussed "another method of attack"
on the question of the meaning of artificial transmutation. A close look at the
atomic masses of hydrogen and helium shows that "in the synthesis of the
helium nucleus from hydrogen nuclei . . . a large amount of energy in the
form of radiation has been released in the building of the helium nucleus from
its components."[53] Arthur Eddington had suggested that this source of
energy may account for the heat emission of sun and stars. Before the same
forum in 1920 Eddington had speculated that the interior of stars is the place
where the evolution of all the elements from hydrogen occurs, and, provoca-
tively, if "the sub-atomic energy in the stars is being freely used to maintain
their great furnaces, it seems to bring a little nearer to fulfillment our dream
of controlling this latent power for the well-being of the human race — or

for its suicide."[54] Rutherford agreed that the evidence of stellar evolution "certainly indicates that the synthesis of helium, and perhaps other elements of higher atomic weight, may take place slowly in the interior of hot stars."[55] The facilities of the Mount Wilson Observatory, along with those of the Norman Bridge Laboratory in Pasadena, seemed to place Millikan in a strategic position to investigate these exciting problems.

At this point, around 1922, Millikan welded together his two interests. Before them, Millikan's upper-atmosphere research program emphasized *testing* the existence of rays from space; only in 1922 did he make explicit his view that the penetrating rays might have their origin in the cosmic processes so interesting to the astrophysicists. The earliest evidence yet uncovered wherein Millikan makes this connection is his report "Fundamental Researches on the Structure of Matter" for the Carnegie Institution of Washington in 1922:

These penetrating radiations *must apparently have their origins in nuclear changes going on in the atoms of the sun and stars*, and their study is therefore a very fitting part of the program for the joint attack on the problem of the structure of matter from both the physical and the astrophysical points of view.[56]

Moreover, he began by 1923 to see the penetrating radiation as a link between the nuclear transformations of the elements in the heavens and those occurring on earth. A paper he presented before the Carnegie Institution, "Atomic Structure and Etherial Radiation," made this link explicit. Writing to J. C. Merriam of the Carnegie Institution on May 3, 1923, Millikan indicated that he chose this topic "because it illustrates the beauty, and the consistency too, of our vision, a vision which has been acquired within the past two decades into the structure of matter, and at the same time the abysmal depths of our ignorance when we attempt to go a little farther and relate these different fields of physical investigation sufficiently to a consistent whole."[57]

In the address itself Millikan noted that when "radioactivity was first discovered it was conjectured by some that the energy involved in this radioactive change might come from outside somewhere."[58] Focusing specifically on the energy for beta decay, Millikan posed the important question, "Where does the energy come from which enables the negative electron to push itself out against the pull of the nucleus in which it certainly lies ... ?" In reply he offered a possible "way out": "I have been interested in recent years because of that difficulty in attempting to find out whether there are any penetrating radiations that come in from the outside."[59]

The notion to which he referred, that the energy for radioactivity may come from "outside somewhere," had first been raised by Marie Curie in 1898. Mme. Curie suggested that "all space is constantly traversed by rays analogous to Röntgen rays but which are much more penetrating and which can be absorbed only by certain elements of heavy atomic weight such as uranium and thorium."[60] The idea was revived with considerable éclat by Jean Perrin in 1919. Perrin pointed to the known existence of highly penetrating radiation widely viewed as extraterrestrial in origin and suggested that these "rayons ultra X" were responsible for radioactive dissociation.[61] His views created, somewhat later, a flurry of experimental activity, especially after Millikan's work on cosmic rays renewed interest in this area.[62]

During the winter of 1921–1922, with the assistance of Ira Bowen and head mechanician Julius Pearson, Millikan constructed self-recording instruments for his ballon flights. Each balloon would carry a barometer, a thermometer, and an electroscope which along with the accompanying photographic film weighed only 198 grams. Bowen and Millikan sent their equipment aloft from Kelly Field, Texas, in flights that attained altitudes of over 15 kilometers. The results must have been disappointing. They did not, at least, absolutely confirm the existence of penetrating radiation from above. Prevailing views would lead them to expect to find very high rates of discharge, because only about 12 percent of the atmosphere was left to absorb the radiation. They did not, Millikan reported, "find anything like the computed rates of discharge."[63] They were able to show, however, that if the rays came from above, they were far more penetrating than had been supposed.

Further tests were made in 1922 and 1923 by Millikan and Russell Otis from atop Mount Whitney and Pike's Peak, but the results were generally agreed to be likewise disappointing and inconclusive.[64] In the *Physical Review* for 1924 Otis and Millikan were able only to state that if one assumed Werner Kolhörster's 1923 conclusions – a penetrating radiation of cosmic origin producing 2 ions per cubic centimeter per second and having an absorption coefficient of 2.5×10^{-3} per centimeter in water – the measured ionization was far too low. Moreover, a local storm significantly altered their results. Their conclusions mark a stunning reversal: "We conclude therefore that there exists no such penetrating radiation as we have assumed . . . [and] that the whole of the penetrating radiation is of local origin. How such quantities of radioactive material get into the upper air is as yet unknown."[65]

This brief note contains a big surprise: the apparent willingness of Millikan in 1924 to revert to a theory of local radioactive elements as the origin of the penetrating radiation. In June 1924 Millikan lectured at University College,

London and discussed at length "The Penetrating Radiations of the Upper Air." The report of the talk in *Nature* confirms the brief *Physical Review* note:

We conclude, therefore, that there can exist no such penetrating radiation as we have assumed [i.e., with Kolhörster's postulated properties]. Our observations therefore seem to us to show that the whole of the penetrating radiation on top of Pike's Peak is of local origin. We have computed its absorption coefficient and find it but a little harder than that of radioactive materials. *How such quantities of radioactive material get into the upper atmosphere is as yet unknown.* [66]

Millikan's skepticism concerning the existence of extra-atmospheric penetrating radiation received support from G. Hoffmann in 1925, who likewise saw it arising from known radioelements,[67] and from F. A. Lindemann.[68] Millikan's report to the Carnegie Institution in 1924 rather cryptically referred to the Otis and Millikan results as appearing "to require a profound modification of previous views as to the origin of the penetrating radiation."[69]

In the summer of 1925 Millikan, with his student G. H. Cameron, set out to settle the question definitively. They traveled to two California lakes, Muir Lake (altitude 12,000 feet) and Lake Arrowhead (altitude 5000 feet). They found that the intensity of ionization demonstrated that the rays, coming exclusively from above, had 18 times the penetrating power of the hardest known gamma rays. Sinking their instruments in the upper lake to a depth of 6 feet (the water-equivalent in absorbing power of the layer of air between the surfaces of the two lakes) they showed that the readings of the lower lake were then identical with that of the upper. They concluded that the atmosphere contributed nothing to the intensity of ionization at the surface of the lower lake. The atmosphere acted only as an "absorbing blanket." The rays came, they said, not from the earth or the atmosphere, but from space.[70] Millikan's skepticism evaporated.

Before the National Academy of Sciences at Madison, Wisconsin, on November 9, 1925. Millikan exuberantly reported upon his new findings and dubbed the highly penetrating radiation "cosmic rays."[71] He was quick to return to his earlier cosmic hypotheses. Since the most penetrating rays producible on earth are from radioactive transformations, he reasoned that

It is scarcely possible, then, to avoid the conclusion that these still more penetrating rays which we have been studying are produced similarly by nuclear transformations of some sort. . . . We can scarcely avoid the conclusion, then, that nuclear changes having an energy value perhaps fifty times as great as the energy changes involved in observed radioactive processes are taking place all through space, and that signals of these changes are being sent to us in these high-frequency rays.[72]

Millikan pointed out that the frequency of the hardest cosmic rays known at that time corresponded to the energy of formation of helium out of hydrogen, and corresponded closely also to the capture of an electron by a light nucleus. Such nuclear captures may in fact be the likely explanations for the origin of such rays. He was very excited about this evidence for what he saw as the "birth cries" of infant atoms, born either by fusion or by electron capture.[73] Before the American Philosophical Society he exclaimed that we can if we wish "call it the music of the spheres!"[74]

The National Academy address excited public attention as little else had previously done in American science. Millikan became, virtually overnight, a media figure. The *New York Times* exulted in an editorial: "Dr R. A. Millikan has gone out beyond our highest atmosphere in search for the cause of a radiation mysteriously disturbing the electroscopes of the physicists," and through "patient, adventuring observations" has "found wild rays more powerful and penetrating than any that have been domesticated." The *Times* went on to insist that these "Millikan rays," as they dubbed them, ought "to find a place in our planetary scientific directory all the more because they would be associated with a man of such fine and modest personality."[75]

Time magazine seized the occasion to exercise its most purple prose:

Dr. Robert Andrews Millikan ... told the Academy about a new ray which had been discovered – a ray which begins in eternity. Born beyond space, in some dim interstellar vestibule behind the gates of the discoverable universe, out of a womb still swollen with gas, perhaps with litters of uncreated stars, the Millikan Ray stabs earthward, traversing aerial shambles strewn with the débris of mutating solar systems, planes where ... parallel lines may meet, and voids in which time, unhinged, spins like a tiny weathervane in an everlasting whirlwind.[76]

Within a year and a half, Millikan, inexplicably peering through a microscope, graced the cover of *Time*, which breathlessly exclaimed that he had "detected the pulse of the universe.[77]

When Victor Hess expressed chagrin at the use of the term "Millikan Rays" and dismay at the popular success that Millikan was enjoying as the "discoverer" of penetrating or cosmic rays,[78] Millikan wrote a letter to him expressing regret that the term was employed, and assuring him that

The really important thing is that between all of us we have been able to make pretty certain the existence of a radiation which comes to earth from outside. The evidence for this origin ... has not been convincing to a great many physicists including Swann and myself in this country, Lindemann in England and Hoffman in Germany. The evidence seems to me now to be unambiguous. That such cosmic rays, if they exist, must be of nuclear origin is altogether obvious. It has been suggested literally scores of times.[79]

Millikan's embarrassment, while genuine, was undoubtedly mixed with considerable pride and satisfaction in the public acclaim.

In the *Physical Review* for November 1926 Millikan returned with renewed confidence to the problem of the origin of cosmic rays: "It is altogether obvious that any rays of the hardness and distribution indicated, and of cosmic origin, must arise from nuclear changes of some sort going on all about the earth," but far more energetic than any radioactive change thus far on record.[80] The cosmic rays most probably come from among the following nuclear changes: "(1) the capture of an electron by the nucleus of a light atom, (2) the formation of helium out of hydrogen, or (3) some new type of nuclear change, such as the condensation of radiation into atoms." In any case the changes are not processes occurring in stars but rather, as W. D. Macmillan of Chicago had suggested, are occurring "in the nebulous matter in space, i.e., throughout the depths of the universe."[81] Macmillan earlier in the year had interpreted Millikan's cosmic rays as evidence of "a striking confirmation" of his hypothesis that atoms are being continuously evolved in interstellar space, a view he had discussed with Millikan as early as 1915.[82] "The rays must come," he reaffirmed in early 1928, "in the main from beyond the Milky Way."

By the beginning of March 1928 Millikan was eager to announce another headline-snatching breakthrough. Writing to his son Glenn on February 27 he told a remarkable story:

Night before last about 12:30 I got some new results which gave me quite a fever. Cameron and I had laboriously analyzed our cosmic ray curve and concluded it had to be produced not by general radiation but by bands and we had located three of these bands. These three bands were found Saturday night to fall *just where they should* in the frequency range if they were produced by (1) the formation of helium out of hydrogen, (2) the formation of oxygen out of hydrogen, (3) the formation of silicon out of hydrogen, the three elements which constitute the great bulk of the mass of the earth, of meteorites and of the stars, so far as the latter are amenable to estimates of their constitutions. If these results are valid they constitute the first evidence that the building up as well as the disintegrating process is going on under our eyes, the signal of the birth of an atom of helium, oxygen or silicon being sent out to the ends of the universe wherever such an event occurs in the obstetrical wards of space. . . Maybe it is a group of accidental occurrences, but I doubt it!!![83]

Millikan was astonished and excited. His 1904 article for the *Popular Science Monthly*, 'Recent Discoveries in Radiation and Their Significance,' had posed "a profoundly interesting question": Is there "any natural process which does, among the atoms, what the life process does among the molecules, i.e., which takes the simpler forms and builds them up again into more complex ones?"[84] Now, almost a quarter of a century later he hoped — indeed, he more than half believed — that he had the answer at last. Before the California Institute Association in mid-March he elaborated his view.[85] He analyzed the ionization-depth curves which he and Cameron had laboriously compiled. He indicated that the cosmic ray absorption curve could be accounted for by summing the curves of three sets of presumed frequency bands: a low-frequency band responsible for most of the atmospheric absorption having an absorption coefficient of 0.35, and two high-frequency bands having coefficients of 0.08 and 0.04. Using the precise work of Aston on atomic weights, Einstein's famous relation between mass and energy, and Dirac's recently published formula for absorption through Compton scattering, he was able to compute energies of the cosmic rays (assumed to be light quanta or photons) of about 26, 110, and 220 million electron volts and to show that these energies corresponded to the "mass defects" occurring in the building up, out of hydrogen, of helium, oxygen, and silicon.[86] A report in *Science* by Millikan and Cameron declared the cosmic-ray photons to be "the announcements sent out through the ether of the birth of the elements."[87]

Writing again to his son Glenn at the end of March, Millikan was able to exult:

The philosophy and the science are running together in the new stuff that we have been getting out of cosmic rays ... in a nutshell, the thing that looks good is the numerical agreement between observed absorption coefficients and those computed from [the] Einstein equation giving the interrelationship of energy and mass ($mc^2=E$). The fit seems too good to be a mere coincidence and if it is real it indicates the continuous birth of the ordinary elements out of positive and negative electrons.[88]

Once again the mass media took up Millikan's work as a brilliant illumination of the inner secrets of the universe. The *New York Times'* front page beamed, "Creation Continues, Millikan's Theory," and "Cosmic Rays Herald 'Birth of the Elements.'"[89] Even more dramatically the *Times* insisted that "Super X-Rays Reveal the Secrets of Creation."[90]

When in 1928 the Klein-Nishina formula for relating absorption coefficients with energy replaced Dirac's, Millikan was able successfully to hold fast to his atom-building hypothesis. The revised energy calculations seemed, in fact, to bolster his position.[91] Before the Royal Society, however, Rutherford cautioned that although "the absorption coefficient of the most penetrating radiation deduced by Millikan and Cameron from their experiments is in excellent accord with that to be expected on the Klein-Nishina formula for a quantum of energy 940 million volts, we should be wary of relying on extrapolations of theories of absorption into the other energy ranges," or in other words, physicists had little confidence in the formula at energies required for Millikan's interpretations.[92]

The atom-building hypothesis, which Millikan found so exhilarating, received remarkably little attention from his peers. It generated at first little controversy, and even less support. For Millikan the cosmic photons "must be in fact the birth cries of the elements,"[93] but for others the entire effort seemed contrived. E. C. Stoner, the British physicist, wrote that the probability of hydrogen atoms finding each other in order to coalesce was 'vanishingly small.'"[94]

James Chadwick, soon to discover the neutron, roundly criticized Millikan's atom-building hypothesis and its methodology at a Royal Society "Discussion on Ultra-Penetrating Rays," when he noted that the separation of the radiation into components was "a questionable proceeding, for other interpretations of the absorption curve are possible," and furthermore noted that it is unlikely that the Klein-Nishina formula is valid for the penetrating rays.[95] When during the course of the 1930s the work of A. H. Compton and others demonstrated that cosmic rays were not photons at all, but were in fact composed of charged particles, the underpinnings of the atom-building hypothesis collapsed entirely. Indeed, Millikan and his co-worker Carl Anderson themselves directly measured cosmic-ray energies in 1931 and found them to be far in excess of any that could be accounted for by packing fractions.[96] Millikan began, at this point, a slow retreat to a fall-back position; from atom-building he and his co-workers ultimately passed to atom-annihilation.[97]

But for Millikan during the late 1920s the atom-building hypothesis was a striking confirmation of his faith in an evolving atomic structure, that is, one in which atoms were being constructed as well as radioactively decaying. It was, for him in 1928, a way to avoid the "heat death" or running-down of the universe posed by the second law of thermodynamics,[98] and even if he retreated from this position, at least it was in his eyes "a little bit of *experimental* finger-pointing" toward a Creator who was continually on His job.[99] In short, atom-building was an idea in deep harmony with Millikan's fundamental spiritual yearnings. In 1904, the evolution of the elements, as suggested by radioactivity, was for him a way of rejecting crass materialism. As he later expressed it:

Matter is no longer a mere game of marbles played by blind men. An atom is now an amazingly complicated organism, possessing many interrelated parts and exhibiting many functions and properties — energy properties, radiating properties, wave properties, and other properties quite as mysterious as any that used to masquerade under the name of "mind." Hence the phrases – "All is matter" and "All is mind" – have now become mere shibboleths completely devoid of meaning.[100]

Johns Hopkins University

NOTES

[1] This paper is a summary of part of my book "The Rise of Robert Millikan," (Ithaca: Cornell University Press, 1982). On the origins of the California Institute of Technology, and Millikan's role in it, see Robert Kargon, "Temple to Science: Co-operative Research and the Birth of the California Institute of Technology," *Hist. Stud. Phys. Sci.*, 8 (1977), 3–31.

[2] A useful review is F. H. Loring, *Atomic Theories* (London: Methuen, 1921), pp. 41–68.

[3] Millikan, "Holographic Autobiographical Notes," Robert A. Millikan Papers, California Institute of Technology (henceforth cited as RAM), 67.8. I am grateful to Judith Goodstein, archivist, for permission to quote from them.

[4] On Moseley's work see *Phil. Mag.*, 26 (1912), 1024 and 27 (1914), 1703. See also J. Heilbron, *H. G. J. Moseley: The Life and Letters of an English Physicist, 1887–1915* (Berkeley: University of California Press, 1974), Chapters 5, 6.

[5] H. G. J. Moseley, *Phil. Mag.*, 27 (1914), 703–713. See also R. A. Millikan, *The Electron* (Chicago: University of Chicago Press, 1917), pp. 200–202.

[6] R. A. Millikan, "Radiation and Atomic Structure," *Phys. Rev.*, 10 (1917), 194–195.

[7] *Ibid.*, pp. 204–205.

[8] *Ibid.*, p. 225.

[9] "Memorandum Relating to the Application of the California Institute of Technology to the Carnegie Corporation for Aid in Support of a Project of Research on the Constitution of Matter" (Sept. 17, 1921). George Ellery Hale Papers, California Institute of Technology, Box 6, p. 1.

[10] University of Chicago, President's Report, 1905–1906, p. 124.

[11] T. Lyman, "Spectroscopy of the Ultra Violet," *Astrophys.*, 43 (1916), p. 89.

[12] R. A. Millikan, *Astrophys.*, 52 (1920), 47; R. A. Millikan, R. A. Sawyer, and I. Bowen, *ibid.*, 53 (1921), 150; P. Epstein, "Robert Andrews Millikan as Physicist and Teacher," *Rev. Mod. Phys.*, 20 (1948), 19–21.

[13] R. A. Millikan, *Proc. Nat. Acad. Sci.*, 7 (1921), 289.

[14] R. A. Millikan and B. E. Shackelford, *Phys. Rev.*, 15 (1920), 240.

[15] R. A. Millikan and C. Eyring, *Phys. Rev.*, 27 (1926), 51; R. A. Millikan and C. Lauritsen, *Proc. Nat. Acad. Sci.*, 14 (1928), 45; *Phys. Rev.*, 33 (1929), 598; R. A. Millikan, C. Eyring, and S. S. Mackeown, *ibid.*, 31 (1928), 900.

[16] J. R. Oppenheimer, *Phys. Rev.*, 31 (1928), 914; R. H. Fowler and L. Nordheim, *Proc. Roy. Soc.*, 119 (1928), 173.

[17] C. T. R. Wilson, *Proc. Camb. Phil. Soc.*, 11 (1900), 52; J. Elster and H. Geitel, *Phys.*, 2 (1900–1901), 560.

[18] E. Rutherford and H. Cooke, *Phys. Rev.*, 16 (1903), 183; J. McLennan and E. Burton, *ibid.*, 184.

[19] K. Kurz, *Phys.*, **10** (1909), 834.

[20] K. Bergwitz, *Habilitation-Schriften*, Braunschweig, 1910; A. Gockel, *Phys. Z.*, **11** (1910), 280; *ibid.*, **12** (1911), 595. I am following the account of V. Hess, *The Electrical Conductivity of the Atmosphere and its Causes* (New York: Van Nostrand, 1938), pp. 115–118, and J. D. Stranathan, *The "Particles" of Modern Physics* (Philadlphia: Blakiston, 1942), Chapter 12.

[21] V. Hess, *Ber. Akad. Wiss. Wien*, **120** (1911), 1575; *ibid.*, **122** (1913), 1481. W. Kolhörster, *Phys. Z.*, **14** (1913), 1066; *ibid.*, **1153**; *Deut. Phys. Ges. Verh.*, **16** (1914), 719.

[22] V. Hess, *Ber. Akad. Wiss. Wien*, **121** (1912), 1001; *ibid.*, **122** (1913), 1481-1486.

[23] V. Hess and M. Kofler, *Ber. Akad. Wiss. Wien*, **126** (1917), 1389-1436.

[24] Millikan to Robert Woodward, May 6, 1919, RAM 20.14. See also Millikan to Henry Crew, Mar. 3, 1920, Crew Papers, Niels Bohr Library, American Institute of Physics, New York.

[25] Carnegie Corporation Application, Hale Papers, Box 6, Section D, p. 3.

[26] A. Sommerfeld, *Atombau und Spektrallinien* (3rd ed., Braunschweig: Vieweg, 1922), p. 75.

[27] R. A. Millikan, *The Electron* (Chicago: University of Chicago Press, 1917), pp. 202–203.

[28] F. W. Aston, *Isotopes* (London: Arnold, 1922), p. 90.

[29] *Ibid.*, p. 101; W. Harkins and E. Wilson, "Energy Relations Involved in the Formation of Complex Atoms," *Phil. Mag.*, **30** (1915), 723–734.

[30] Aston, *Isotopes*, p. 104.

[31] E. Rutherford, *Smith. Inst. Ann. Rep.*, 1915 (Washington, D.C., 1916), p. 201.

[32] R. A. Millikan, "Recent Discoveries in Radiation and their Significance," *Pop. Sci. Mon.*, **64** (1904), 49.

[33] *Ibid.*

[34] R. A. Millikan, "Gulliver's Travels in Science," *Scribner's*, **74** (1923), 584.

[35] W. Ramsey, N. Collie, and H. Patterson, *Nature*, **90** (1913), 653–54.

[36] G. Winchester, *Phys. Rev.*, **3** (1914), 294.

[37] E. Rutherford, *The Newer Alchemy* (Cambridge: Cambridge University Press, 1937), passim.

[38] Millikan to Rutherford, Feb. 3, 1920. RAM 42.

[39] R. A. Millikan, "The Significance of Radium," *Science,* **54** (1921), 59–67.

[40] Hale to Henry Robinson, Feb. 25, 1921, Hale, Papers, Box 35.

[41] Hale to James Angell, June 4, 1921, Hale Papers, Box 6.

[42] See Kargon, "Temple to Science," (n. 1 above).

[43] Application to Carnegie Corporation, Hale Papers, Box 6.

[44] Caltech Archives, Institute Publicity Releases, Box 31-A, p. 12.

[45] Millikan, "Gulliver's Travels," p. 584.

[46] He repeated this claim in 1926: "The Last Fifteen Years in Physics," *Proc. Amer. Phil. Soc.*, **65** (1926), 74.

[47] Millikan, "Gulliver's Travels," p. 584. Emphasis supplied.

[48] R. Sorensen, *J.A. I. E.E.*, **44** (1925), 373–374.

[49] G. Winchester, *Phys. Rev.*, **3** (1914), 294.

[50] A. Miethe and H. Stammreich, *Nature*, **114** (1924), 197–198; H. Nagaoka, *ibid.*, **116** (1925), 95–96.

88   ROBERT H. KARGON

51 A. Smits, *Nature*, **117** (1926), 13, 621.

52 F. Haber, *Naturwissenschaften*, **14** (1926), 405–412. F. Aston, *Nature*, **116** (1925), 902–904.

53 E. Rutherford, *BAAS Rep.* (1923), 21.

54 A. Eddington, *BAAS Rep.* (1920), 46.

55 E. Rutherford, *BAAS Rep.* (1923), 21.

56 *Carnegie Inst. Rep.*, **21** (1922), 385–386. Emphasis supplied.

57 Millikan to J. C. Merriam, May 3, 1923, Merriam Papers, Library of Congress, Box 125.

58 R. A. Millikan, "Atomic Structure and Etherial Radiation," RAM 62.10, p. 15.

59 *Ibid.*, p. 16.

60 M. Curie, *Compt. Rend.*, **126** (1898), 1103.

61 J. Perrin, *Ann. Phys.*, **11** (1919), 85–87.

62 See, e.g., L. P. Maxwell, *J. Franklin Inst.*, **207** (1929), 619–628; N. Dobronravov, P. Lukirsky, and V. Pavlov, *Nature*, **123** (1929), 760; L. N. Bogojavlensky, *ibid.*, 872.

63 R. A. Millikan, *Proc. Nat. Acad. Sci.*, **12** (1926), 149.

64 R. A. Millikan and I. Bowen, *Phys. Rev.*, **22** (1923), 198; R. Otis, *ibid.*, 198–199.

65 R. Otis and R. A. Millikan, *Phys. Rev.*, **23** (1924), 778–779.

66 R. A. Millikan, *Nature*, **114** (1924), 143. Emphasis supplied.

67 G. Hoffmann, *Phys. Z.*, **26** (1925), 669–672.

68 Millikan to Hess, July 24, 1926, RAM 40.9.

69 R. A. Millikan, *Carnegie Inst. Rep.*, **23** (1923–24), 301.

70 R. A. Millikan, *Electrons (+ and −), Protons, Photons, Neutrons, Mesotrons and Cosmic Rays* (Chicago: University of Chicago Press, 1947), pp. 307–308.

71 R. A. Millikan, *Proc. Nat. Acad. Sci.*, **12** (1926), 48–55.

72 *Ibid.*, pp. 53–54.

73 R. A. Millikan, *Science and the New Civilization* (New York: Scribner's, 1930), p. 105.

74 R. A. Millikan, "The Last Fifteen Years of Physics," *Proc. Amer. Phil. Soc.*, **65** (1926), 78.

75 *N.Y. Times*, Nov. 12, 1925, p. 24.

76 *Time*, **6** (May 23, 1925), 26–27.

77 *Time*, **9** (Apr. 25, 1927), cover.

78 V. Hess, *Phys. Z.*, (1926), 159–164.

79 Millikan to V. Hess, July 24, 1926. RAM 40.9.

80 R. A. Millikan and H. Cameron, *Phys. Rev.*, **28** (1926), 867.

81 *Ibid.*, p. 868.

82 W. D. Macmillan, "The New Cosmology," *Sci. Amer.*, **134** (1926), 310–311; R. A. Millikan, "History of Research in Cosmic Rays," *Nature*, **126** (1930), 15.

83 Millikan to Glenn Millikan, Feb. 27, 1928. RAM 57.5.

84 R. A. Millikan, *Pop. Sci. Mon.*, **64** (1904), 498.

85 R. A. Millikan and H. Cameron, "Direct Evidence of Atom Building," *Science*, **62** (1928), 401–402.

86 R. A. Millikan and H. Cameron, "The Origin of Cosmic Rays," *Phys. Rev.*, **32** (1928), 533–557; P. Dirac, *Proc. Roy. Soc.*, **A111** (1926), 423; F. Aston, *ibid.*, **115** (1927), 487–514.

87 R. A. Millikan and H. Cameron, *Science*, **62** (1928), 402.

88 Millikan to Glenn Millikan, M ar. 29, 1928. RAM 57.5.
89 *N.Y. Times*, Mar. 18, 1928, p. 1.
90 *N.Y. Times*, Mar. 25, 1928, Sec. 10, p. 3.
91 R. A. Millikan and I. Bowen, *Proc. Nat. Acad. Sci.*, **16** (1930), 423.
92 E. Rutherford, *Proc. Roy. Soc.*, **A122** (1929), 15.
93 Millikan, "The Significance of Cosmic Rays." RAM 63.28.
94 E. C. Stoner, *Leeds Phil. Lit. Soc. Proc.*, **1** (1929), 349. See also Stoner, *Phil. Mag.*, 7 (1929), 841–858.
95 J. Chadwick, *Proc. Roy. Soc.*, **132** (1931), 343.
96 R. A. Millikan and C. Anderson, *Phys. Rev.*, **40** (1932), 325; *ibid.*, **45** (1934), 352.
97 R. A. Millikan, et al., *Phys. Rev.*, **61** (1942), 397–407; *Nature*, **151** (1943), 66.
98 R. A. Millikan and H. Cameron, *Phys. Rev.*, **32** (1928), 555–557.
99 R. A. Millikan, *Nature*, **127** (1931), 170.
100 R. A. Millikan, "What I Believe," *Forum*, **82** (1929), 197–198.

SPENCER R. WEART

THE DISCOVERY OF FISSION AND A NUCLEAR
PHYSICS PARADIGM

How was nuclear fission discovered? Many accounts of this, the most famous discovery in modern physics, are available. Yet when we enter the territory we find that the maps are worse than we might have expected. Some main features have not been mapped at all, or their shape and nature have been poorly explored. And this brings us to the general problem of how to map scientific work. When we study the history of fission we find a type of terrain — a style of doing scientific work — that is highly important yet rarely explored in detail.

Many people have noticed that doing good scientific work usually has much in common with the everyday experience of solving a puzzle. When we look closely into even the greatest upheavals in scientific thought, the growth of ideas like the quantum nature of energy, we find that these ideas grew amidst numerous little technical problems confronted and partly solved, problems like how the light emitted inside a white-hot box is distributed among the colors of the spectrum. Experimentalists and theorists alike spend most of their scientific careers untangling such tight little knots. Yet the essential role of this research is too little appreciated in the history of science, where sweeping theories tend to attract most of our attention.

The history of fission offers no such dazzling theories. If there were no nuclear weapons or reactors, I believe hardly anyone would have studied the history of fission: in terms of science, it comes to our attention almost by accident. In the history of the discovery of fission, then, we have an opportunity to observe vital components of scientific work which in other cases may be overshadowed by spectacular theoretical or experimental upheavals. At the same time, the practical meaning of fission is so great that we dare not be content with unraveling technical details of the history, but must go on to confront more general questions of what drives scientific work.

This study therefore has two main parts. The first two-thirds show what the puzzles were that led to the discovery of fission, how these puzzles were tackled, and how they were finally solved. While not lacking in human elements — rivalries and disappointments and surprises — the story is chiefly about technical points. It was there among the technicalities that the important events took place. The latter third of the study looks back over the

91

William R. Shea (ed.), Otto Hahn and the Rise of Nuclear Physics, 91–133.
Copyright © 1983 *by D. Reidel Publishing Company.*

history for something more general. A set of concepts is uncovered that ran all through the scientific work, a single set that both defined the puzzles of interest and suggested how to solve them. It is found that this set of concepts gives us a concrete and detailed example of a *paradigm*, that is, of the sort of thing that Thomas Kuhn, in his well-known work on the structure of scientific revolutions, claims must underlie any progressive scientific field. I shall not have much to say about scientific revolutions themselves. For the discovery of fission, I repeat, did not involve any wholesale revolution of ideas about nature; it was not physics but industry and warfare that were radically changed. Perhaps this explains why the discovery of fission has hitherto attracted far fewer scholarly studies than popular accounts.[1]

The popular mind would picture the discovery of nuclear fission as taking place in an isolated laboratory room, walled off from the world, with a lonely scientist working late at night, scratching his head over notebooks and instruments. This picture is perfectly accurate for the moment of discovery of fission; yet this moment was only a minor incident in a larger process. Rather than imagine a single room, we should call before our mind's eye Otto Hahn's entire institute in Dahlem, a squarish multi-story building with columns and turrets, a warren of rooms full of advanced students coming and going and working on all sorts of researches. Nearby in this suburb of Berlin we would notice other institutes within the Kaiser Wilhelm Society, so that Hahn, his collaborator Lise Meitner, and their students are found embedded in a larger network of scientific discussion. Like all good institutes, Hahn's had its traditions for getting people together in good spirits, not only colloquia but merry parties on Christmas Eve, birthday celebrations, summer excursions, and every afternoon a rendezvous over tea. People from other institutes in Berlin-Dahlem were often present. Beyond this the institute was connected to a wider scientific world. Every week the mail brought a stack of correspondence as well as periodicals with printed communications that often had been written only a few weeks earlier. Hahn and Meitner themselves published their results about four times a year. They also took part in the floating seminar of nuclear physics, conferences and summer schools constantly dispersing and forming again at a new location. During the 1930s Hahn or Meitner would be off to Copenhagen, Zurich, London, Rome, New York, and Moscow, discussing the latest developments with their peers. Their work becomes intelligible only if we see it less as an isolated task than as one manifestation of the general ferment in nuclear physics. In particular, the ideas behind the discovery of fission — the ideas that delayed the discovery as well as those that finally made it possible — turn out to be intelligible only in terms of the general thought of nuclear physics of the 1930s.[2]

In the early 1930s many nuclear physicists and chemists were involved in experimental work, all of which fell into much the same pattern. One took some substance, bombarded it with particles (from a naturally radioactive element or from a particle accelerator), and studied what came out. In this fashion bombardment of beryllium with alpha particles was found to produce neutrons, and a little later bombardment of aluminum with alpha particles was found to induce artificial radioactivity. Soon after, in 1934, Enrico Fermi and his colleagues in Rome combined the two processes, bombarding elements with neutrons in order to make the elements radioactive. His team found that the neutron could easily slip into an atom and be absorbed by the nucleus, making an isotope heavier by one atomic mass unit. The newly constructed nucleus would usually emit a beta particle and thereby transmute to a more stable isotope, with an atomic number one unit higher.

Fermi's team marched through many of the elements, irradiating each in turn with neutrons. At length they came to the heaviest known element, uranium. As expected, neutrons changed some of the uranium atoms into new radioactive substances. The team applied some rather crude chemical tests and concluded that the new substances did not behave chemically like any of the known heavy elements. The new substances were not anything from uranium (atomic number 92) down through radium (number 88). Fermi's team therefore thought it likely that they had created one or more unknown new elements, heavier than uranium.

The discovery of an element has traditionally been a matter for national pride. The creation of an entirely new element could hardly be less, and in Mussolini's Italy the discovery was cause for loud celebration. Fermi was sorely embarrassed, for he was not yet sure enough of the existence of the new elements to stake his reputation on them.[3]

Indeed, doubts about the discovery were soon voiced. Aristide Grosse questioned whether the Rome team had really found transuranium elements. He reported experiments suggesting that Fermi might just as easily have been observing a new radioactive isotope of protactinium, a known element (atomic number 91). This paper attracted the attention of Hahn and Meitner. Not only was the scientifically interesting and prestige-laden question of new elements at stake, but protactinium was a specialty of theirs, for they had shared its discovery in 1917. More recently, starting in 1931, they had gotten into an unpleasant controversy with Grosse about protactinium in the pages of *Die Naturwissenschaften*. Now they decided to find out whether or not he was right this time.[4]

The methods Hahn and Meitner now used were well-tried techniques that would characterize much of their later work on uranium. First, of course, they irradiated uranium with neutrons. Next they undertook chemical tests of the products, dissolved in an acidic solution. The new elements were produced in minuscule amounts by the neutrons, sometimes only a few thousand atoms after hours of irradiation. These atoms were detectable only by their radioactivity. There was no way to precipitate such a tiny amount of stuff, so it was customary for radiochemists to add a similar element as a "carrier"; the radioactive atoms would be entrained and follow the carrier through the operations. For example, to test for an element heavier than uranium that they expected would behave chemically like rhenium, Hahn and Meitner added a rhenium compound to their solution, then introduced hydrogen sulfide and precipitated out rhenium sulfide. The radioactivity tended to precipitate out along with the rhenium. This showed that the radioactive atoms were chemically similar to rhenium and likely to be the transuranic "eka-rhenium" they were looking for. A similar test with platinum worked even better, proving that Grosse was wrong: the radioactive elements, whatever they were, did not behave like protactinium.

Hahn and Meitner noticed that precipitation with hydrogen sulfide was a remarkably good way to separate out transuranium elements. In particular, the method would exclude the natural decay products of uranium, which otherwise would plague their experiments. Radiochemists who worked with uranium normally purged it at the outset of its associated radium and protactinium. But the ordinary, endless decay of uranium would swiftly build up these daughter products again, and within a few hours the newly created radium and protactinium would be abundant enough to swamp the radioactivity of anything else that was present. Using hydrogen sulfide precipitation after they irradiated the uranium, Hahn and Meitner could leave these contaminants behind in solution; then they would be able to detect the feeble radioactivity that might point to transuranium elements. Furthermore, not only protactinium and radium but all the other elements down at least as far as polonium (number 84) could be expected to stay behind in the solution. Hahn and Meitner therefore thought it "very probable" that the radioactive substances in their precipitate were indeed new elements beyond uranium.[5]

Another challenge to Fermi's results, independent of Grosse's, had been published by the German chemist Ida Noddack. She too had noted that

Fermi's chemical tests were not conclusive in eliminating all elements slightly lighter than uranium; she pointed particularly to polonium. She also remarked that many other, much lighter elements were not chemically excluded. Since the effects of neutron bombardment were still poorly understood, she said, it was conceivable that an atom could be broken apart into "several large fragments, which could be indeed isotopes of known elements, but not neighbors of the irradiated elements."[6]

Noddack's paper was read by the scientists concerned, but they paid it little attention. Science does not lack for comments and speculations, and it seemed that Noddack's main point was not the odd passing remark about a nucleus breaking apart, but the reiteration that Fermi's tests had not excluded known elements such as polonium. The subsequent work by Hahn and Meitner, and also by Fermi's group, squelched this objection, eventually removing from consideration all elements down to mercury (number 80). Scientists who reviewed the uranium researches wrote their articles as if all objections, Grosse's and Noddack's together, were answered.[7] They all had a blind spot, which we will have to return to later on. For as we now know, most of the new radioactive substances were indeed elements far lighter than uranium. Fermi had split the uranium nucleus in twain, but years would pass before anyone would understand this.

Hahn and Meitner did recognize that they had a puzzle on their hands. For they found that their precipitate contained more than one type of radioactive substance. By the spring of 1935 both Fermi's team and Hahn and Meitner were reporting substances with three different half-lives. Both groups hesitated to say how these were related and recognized that irradiating uranium produced interesting problems.[8] Fermi's collaborators were now dispersing, but Hahn and Meitner decided to stay together and unravel the puzzle.

Why did Hahn and Meitner devote themselves to uranium? The field of radioactivity included a host of questions, and we may ask what made this one especially attractive. Over the last several years Hahn had been working in applied radiochemistry, using radioactive tracers. His studies were not applied in the direct sense that they could be used immediately be industrialists, but they were steps in this direction. For example, in the fall of 1934 he was studying the crystallization of iron oxides, materials which, he pointed out, were extremely important for heavy industry. His research was therefore well suited to the Kaiser Wilhelm Society, most of whose institutes were devoted to applied science; his own Institute for Chemistry devoted itself to many practical problems and was funded largely from industrial sources. Further, Hahn thought applied radiochemistry a highly promising field for research.[9] He could have continued in this way indefinitely. So it is reasonable to believe Meitner's statement that it was she who approached Hahn over uranium, suggesting they resume their old collaborative search for radioelements after a lapse of nearly fifteen years. In the last few years Meitner's work had produced no very exciting results, so when Fermi's spectacular discoveries were announced she was ready to shift to a new field. She was so impressed by artifical radioactivity that she began neutron bombardments by herself; the role of Grosse's paper was probably just to turn her attention and Hahn's to the specific element uranium.[10]

More than scientific excitement was at stake. At this time Meitner and Hahn were witnesses to the Nazi takeover of German society and the brutal expulsion of German Jews. Although protected from expulsion by her Austrian citizenship, Meitner was anxious, and Hahn was in an uneasy position not only on account of his old Jewish friends but also through his political views, known to be anti-Nazi. It is likely, then, that Hahn and Meitner's decision to work on uranium was influenced by hopes for greater security through enhanced prestige, such as the discovery of new processes or new elements might bring. (We shall see presently how Hahn told Ernest Rutherford about their need for international recognition.) Meitner and Hahn probably felt also that the discovery of new elements would lead not only to scientific gains, not only to prestige, but also in the long run to practical applications. These would be applications in biomedicine or radio-chemical studies; Hahn and Meitner did not mention the possibility of winning the energy of the nucleus. Their views were probably similar to those held by their colleague Carl Friedrich von Weizsäcker, who said that nuclear energy was indeed a possibility, but one not likely to be realized soon.[11]

Proceeding with their irradiations of uranium, Hahn and Meitner found more and more substances with more and more distinct half-lives. They laboriously classified these substances according to chemical behavior. To catch the shortest periods, down to less than a minute, they had to do chemical separations and radioactivity counts at breakneck speed, while for longer half-lives, stretching dozens of hours, they or their student assistants had to take turns watching the counters day and night. Gradually they learned to distinguish which substance was produced from which, but an overall explanation eluded them.

To make their task harder, nobody knew just what the chemistry of transuranium elements should look like. The most straightforward guess was that the periodic table could simply be extended downward: if uranium was regarded as lying beneath the similar metallic element tungsten, then the first element beyond uranium, number 93, should lie beneath the element next to tungsten, namely rhenium. After this eka-rhenium would come eka-osmium, eka-iridium, etc. However, people soon pointed out the danger of encountering a transition group, like the rare-earth group that begins with lanthanum. The elements in such a group would all have almost identical chemical properties. Such a group might begin at atomic number 90 (thorium), or at uranium, or at some other nearby element. Until the transuranium elements were definitively identified and their chemistry worked out there was no way of telling for sure where they would fit into the periodic table, what they would behave like.[12]

To compound the confusion, the decay scheme for the new substances did not resemble the familiar chains of decays that natural radioactive series follow. Irradiated uranium produced complex products. As Hahn wrote Rutherford, "although we have definitely established three disintegration processes for Uranium and believe we have determined the chemical properties of the artificial atomic types with considerable accuracy, we are not yet certain of the correct interpretation of these processes."[13]

Of all the properties of the set of new substances, two were particularly strange. The first was that each element decayed into its successor through a beta emission, one after another, so that the atomic number kept increasing, building as it seemed from 93 up to 96 and beyond. Nothing like this happened in any natural radioactive series, where alpha-particle emissions are commoner than beta decays. When Meitner was interviewed in 1963 about her radium work, this was the puzzle that she immediately recalled: "How can the nuclear charge really increase with the mass held constant?" She recalled that she kept bringing up the problem with von Weizsäcker, a Berlin theorist who often showed up at the afternoon teas in Hahn's institute.[14] However, with nuclear theory in a primitive state and the experiments all uncertain, the problem was not urgent.

A second peculiarity, more important in the long run, emerged as Meitner and Hahn, with the collaboration of Fritz Strassmann, found more and more half-lives in their precipitate. These substances were gradually disentangled and found to be linked genetically in three chains, each one starting with a uranium nucleus irradiated by neutrons. The team found these chains difficult to interpret.

To be sure, three possible outcomes were known when a neutron struck a nucleus. The neutron might provoke the emission of an alpha particle; it might provoke the emission of a proton; or the neutron might simply be captured. In each case the resulting nucleus would usually then emit a beta particle. These three sorts of outcomes were known, for example, to occur in the prototypical case of artificial radioactivity — aluminum. The Berlin team displayed this example for the benefit of their chemist readers when they reviewed their work in a chemical journal in the spring of 1936:

(1) $Al_{13}^{27} + n_0^1 \rightarrow Na_{11}^{24} + He_2^4$; $Na_{11}^{24} \xrightarrow{\beta} Mg_{12}^{24}$

(2) $Al_{13}^{27} + n_0^1 \rightarrow Mg_{12}^{27} + H_1^1$; $Mg_{12}^{27} \xrightarrow{\beta} Al_{13}^{27}$

(3) $Al_{13}^{27} + n_0^1 \rightarrow Al_{13}^{28} \xrightarrow{\beta} Si_{14}^{28}$

Hahn and Meitner identified one of their chains as type 1, with neutron bombardment of uranium yielding an alpha particle plus thorium, the thorium then transmuting through a string of beta decays. Another of their chains seemed to be of type 3, simple neutron capture producing uranium-239 and then its decay products. But the last of their chains did not seem to fit any of the three expected processes. The Berlin team was willing, therefore, to go a little beyond the known processes. This was the first of a number of steps, each bolder than the last, by which scientists were to use the results of uranium research in an attempt to improve their understanding of how nuclei behave in general.

In this case, Meitner and Hahn supposed that the incoming neutron chipped another neutron off the uranium nucleus, with both neutrons flying away. The result would be a uranium-237 nucleus, which could subsequently undergo its own particular set of beta decays. The Berlin team recognized that they were claiming the discovery of an entirely new type of nuclear reaction.[15]

But this was not a large leap. It was intuitively plausible that a neutron could be chipped off, just like a proton or alpha particle. Indeed, at one point Fermi's group had suggested that this could explain some of their results. Within a few months of the Berlin group's publication, though perhaps independent of it, a Russian physicist found the process did occur beyond doubt in beryllium. In October a Dutch physicist demonstrated it in copper and zinc, and within a year there was evidence for over thirty isotopes that would yield two neutrons when bombarded with one.[16]

This was no help to the Berlin team, however, who by late 1936 had abandoned this explanation for their uranium results. For it was clear from the energies involved that only a fast neutron could chip off another neutron — and they found that the uranium decay chain in question could be initiated not only with fast neutrons but with slow neutrons too. Further physical experiments, and difficult chemical tests, finally convinced Meitner and Hahn that this chain began with the simple capture of a neutron. But now they had two chains that began this way, that is, with the creation of uranium-239. And how could one isotope give rise to two different sets of beta decays, each set with its own characteristic half-lives? It was, Hahn admitted, a serious puzzle.[17]

Meitner saw only one way out: to boldly assert the existence of a new property of nuclei. It had long been suspected that a nucleus might have "isomers," that is, states with the same mass and charge yet with different types of radioactivity. Hahn himself had reported such a case in 1921. His "uranium Z" was an isotope of protactinium that was apparently identical to the familiar isotope ("uranium X_2") in every respect except half-life. But uranium Z remained a single, isolated example; it was not sufficient by itself to convince most physicists that isomers could exist. Textbooks tended either to avoid mentioning uranium Z altogether, or to avoid mentioning that it appeared to have the same mass as uranium X_2. In 1935 Irène Curie speculated that uranium Z might be two mass units lighter than uranium X_2. In 1936 Franco Rasetti called uranium Z a "peculiar phenomenon" and said that if this "apparent branching" was really true, then "there must be effective an unknown prohibition mechanism" preventing uranium Z from decaying as fast as uranium X_2. Von Weizäcker, in a theoretical textbook published in 1936, declared that isomers "in fact do not seem to occur," for unless more examples were found, one must assume that more experimental work on uranium Z was needed.[18]

However, neutron bombardment was producing more and more radioactive isotopes, and other anomalous cases were cropping up. For example, in 1935 Leo Szilard and T. A. Chalmers reported that it seemed to be possible to induce two kinds of radioactivity, two different half-lives, in a single isotope of indium. Could not the multiple half-lives that Hahn and Meitner had induced in uranium be similarly explained? Meitner and Hahn fought out the pros and cons of isomers in lively discussions. In the summer of 1936 Meitner summarized the evidence for isomerism in uranium, and pointed out similar problems with indium and a couple of other elements. Her case was convincing.[19]

In November von Weizsäcker produced a theoretical explanation. An isotope could be caught in a metastable excited state, from which emission might be much slower than from the ground state. Specifically, if a nucleus had two states of about the same energy but with very different spins, the states could have quite different half-lives. Further examples of isomerism were soon found by experimenters, and the phenomenon joined the repertory of known nuclear processes.[20] Thus the Berlin group's picture of transuranium elements, at just the point where it had seemed most threatened, had not only found an explanation but had made a valuable contribution to the general understanding of nuclear physics.

Further confirmation of their picture came when Meitner, Hahn, and Strassmann made a close study of their third decay chain. This turned out to start with the "resonance" absorption process, in which a neutron of a specific energy is absorbed. Such resonant captures had been studied closely in various laboratories in the mid-1930s and were relatively well understood. The Berlin team found they could measure the resonant energy. The product, which had a half-life of 23 minutes, also conformed to expectation. For example, the team made a series of precipitations. The amount of activity with a 23-minute half-life was always in proportion to the weight of the uranium precipitated; thus there was little doubt that the 23-minute substance was indeed uranium. Since it then underwent beta decay, raising its atomic number by one unit, one could scarcely deny that the result was a transuranium element. This series of experiments seemed flawless — and so it was, for the team was indeed observing the production of the first transuranium element, later called neptunium. Only the inevitable buildup of radium and protactinium in their solution, swiftly masking the feeble radioactivity of the uranium-239 and neptunium, kept the team from pressing on to more complete chemical studies of their new element.[21] The unimpeachable identification of the resonant capture, like the confirmation that isomerism exists, lent a deceptive appearance of solidity to the overall structure of the Berlin team's research.

To make their work even more irreproachable, the team had to prove that the elements in question were different from the carrier elements (iridium, platinum, etc.) that they always used in order to get enough stuff to handle. The Berlin team checked this and found they could indeed make the separation. For example, when they heated a mixture they found that some of their radioactive elements sublimated away more readily than the carriers. It never occurred to Hahn, Strassmann, or Meitner that instead of being heavier than the carriers and lying in the bottom row of the periodic table, these substances were yet lighter elements one row *above* the carriers.[22]

To summarize the results so far, not only Meitner and Hahn but also a number of other scientists discussing their work used a diagram like that in Figure 12. At the bottom is the resonance process, with its 23-minute decay

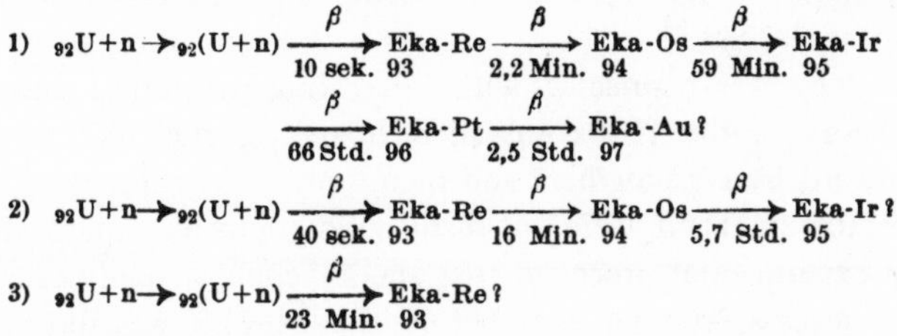

$$1) \quad _{92}U+n \rightarrow _{92}(U+n) \xrightarrow[\text{10 sek. 93}]{\beta} Eka\text{-}Re \xrightarrow[\text{2,2 Min. 94}]{\beta} Eka\text{-}Os \xrightarrow[\text{59 Min. 95}]{\beta} Eka\text{-}Ir$$

$$\xrightarrow[\text{66 Std. 96}]{\beta} Eka\text{-}Pt \xrightarrow[\text{2,5 Std. 97}]{\beta} Eka\text{-}Au\,?$$

$$2) \quad _{92}U+n \rightarrow _{92}(U+n) \xrightarrow[\text{40 sek. 93}]{\beta} Eka\text{-}Re \xrightarrow[\text{16 Min. 94}]{\beta} Eka\text{-}Os \xrightarrow[\text{5,7 Std. 95}]{\beta} Eka\text{-}Ir\,?$$

$$3) \quad _{92}U+n \rightarrow _{92}(U+n) \xrightarrow[\text{23 Min. 93}]{\beta} Eka\text{-}Re\,?$$

Fig. 12. The elaborate, largely mistaken description of neutron reactions with uranium developed in Berlin. Lise Meitner, *Scientia*, International Review of Scientific Synthesis (Via Guastalla 9, Milano, Italy), Annus XXXII, 1938, p. 13.

into element 93, "eka-rhenium." Above are the two other processes with their long beta-decay chains. The general features of this scheme and some of the particular half-lives were publicly confirmed in 1937 by Irène Curie in Paris and, independently, by Philip Abelson in Berkeley and by a group at the University of Michigan, all of whom tried their hand at bombarding uranium with neutrons.[23] By 1938, then, the nuclear physics community had little doubt about the existence of the transuranium elements, and in textbooks, review articles, public lectures, and the like these elements were treated as established fact. At the end of 1938 Fermi received a Nobel Prize for "demonstrations of the existence of new radioactive elements," and in his Nobel lecture he announced the names already in use in Rome for two of the elements, "Ausonium" and "Hesperium."[24]

But when it came to the details of the Berlin group's scheme, misgivings were starting to emerge. It was the Berlin group itself that noted the worst difficulty. Because each of the three processes in Figure 12 was reasonably strong, they believed that each must begin with the common isotope of uranium, namely uranium-238. Then after addition of a neutron, each beta-decay chain would begin with uranium-239. This isotope therefore appeared to have not two but three isomers. And if isomerism consisted, as von Weizsäcker proposed, of one metastable state and the ground state of a nucleus, it was hard to imagine where a third isomer could fit in.

At first the Berlin team had expected that one of their processes began not with a simple neutron capture but with emission of an alpha particle. As before, they searched for alpha-particle emissions, and once again found none. Meitner, Hahn, and Strassmann concluded that the von Weizsäcker model of isomerism must be reconsidered. In previous papers the Berlin group had declined to speculate, but now they offered a tentative suggestion: perhaps isomerism could be explained through some sort of analogy with the construction of molecules.[25]

This analogy was connected with interesting theoretical developments. Nuclear theory in the 1930s was an inchoate, agitated blob of ideas that inched forward by grasping here and there onto an experimental result or a plausible equation. There were not many solid points of contact with reality, not many experimental processes that theory could actually calculate. The first great success, coming even before the neutron was discovered, was George Gamow's theory of alpha-particle decay. He imagined the nucleus as nothing more than a potential well with particles inside – a bag, through whose skin an alpha particle might sooner or later tunnel. This first, and for some time only, successful explanation of a nuclear process profoundly impressed physicists. Yet it tended to inhibit the way of thinking that could lead toward understanding that a nucleus can split in half. From the outset Gamow's theory made it clear that nuclear fragments even slightly larger than an alpha particle would have a hard time slipping through the barrier. An alpha particle had a charge of only two units, and Gamow showed that a fragment with a charge of, say, six units could never be seen to tunnel out. This picture was firmly embedded in the minds of nuclear physicists, including in particular the Berlin group. It was one reason they assumed that a uranium nucleus could never find a way to transform in one step into a substantially lighter nucleus: how could a substantial fragment escape?[26]

Experimental practice reproduced in hardware the theorists' conviction that no particle heavier than an alpha could escape. In several laboratories searches were carried out for energetic alpha particles emitted after uranium or thorium was bombarded with neutrons. To screen out the well-known weak alpha particles emitted by uranium and thorium themselves, the experimenters habitually put a thin foil in front of their detectors. It seemed irrelevant that this foil would also block any products, no matter how energetic, that were substantially larger than alpha particles; for there was no suggestion that such larger particles could be present. Had this practice of using foil been set aside, the very energetic, massive fragments of nuclear fission would surely have been noticed in one or another laboratory. Whether they would have been understood is another question: a group in Zurich did in fact observe something, but concluded that they were unusually energetic alpha particles and failed to publish further.[27]

Despite Gamow's success in explaining alpha-particle emission, nuclear theory was thoroughly confused until the discovery of the neutron was announced by James Chadwick in 1932. Then theorists, following Werner

Heisenberg, could fill up the nuclear potential well with neutrons and protons, and proceed to consider these particles' interaction with the potential. (For more on all these matters, see Roger Stuewer's chapter in this volume.) The simplest approach, considering one particle at a time interacting with a nuclear force, did not readily yield solutions, however. By 1936 Gregory Breit, Eugene Wigner, and Niels Bohr had sketched out a more sophisticated "compound nucleus" model. As Rutherford explained to Chadwick, Bohr proclaimed that "it is impossible to deal with the movements of an actual particle within a nucleus. He therefore considers that the space inside the potential barrier is a sort of 'mush'."[28] A number of theorists began trying to understand this mush, this collection of particles all interacting with one another at the same time. In its early formulation, the compound nucleus model was no better than Gamow's in giving hints of nuclear fission. One still imagined the nucleus as a unit, a bag from which a particle escaped, with the escape of an alpha particle vastly more likely than the escape of a larger fragment.[29]

Meitner and her colleagues tended to think not in terms of the pure compound nucleus model, but of some sort of hybrid picture in which the nuclear particles neither behaved as individuals nor were homogenized into a collective mush. Many physicists noted that since alpha particles were especially stable, the neutrons and protons in a nucleus should tend to clump together to form alpha particles. This picture conveniently explained the emission of an alpha particle when a neutron struck a heavy nucleus: the neutron simply gave up its energy to a pre-existing component of the nucleus, which now had enough energy to escape.[30]

A particularly detailed alpha-particle model was developed by Wilfrid Wefelmeier, a student in Berlin-Dahlem who frequently discussed the problem with von Weizsäcker. Wefelmeier imagined that nuclei were made of alpha particles stacked together geometrically like a pile of spheres, as a molecule is built up of atoms. He proposed that the results of Hahn, Meitner, and Strassmann's work, and particularly the tendency of their radioactive substances to emit one beta particle after another, could be understood if heavy nuclei were stacked in a somewhat nonspherical lump, exposing more surface than a spherical shape. This image of a "nuclear sausage" (*Kernwurst*), as the Berlin physicists dubbed it, eventually showed von Weizsäcker that it is possible for a nucleus to stretch so far that it can divide in two — but he realized this only in 1939, after fission was signalled by experiment.[31]

Another route toward theoretical understanding of fission was the development, following Gamow, of a model in which the compound nucleus was seen as a sort of liquid drop, complete with a temperature, a surface tension, and classical vibrations. By 1938 John Wheeler and a student of his at Princeton were studying the behavior of highly charged, spinning liquid-drop nuclei. They found signs of an instability that they could not explain. This was exactly the sausage-like stretching that leads to fission, but their theory had various difficulties and they did not follow up the clue.[32]

Theory was thus well along toward understanding fission, so that when experiments turned up unmistakable evidence, physicists would be prepared to explain and accept the phenomenon. Yet the theorists were not ingenious enough to point the experimenters straight to fission, an idea outside the bounds of their thinking. Theory served best when it simply suggested that the experiments were getting into something unexplained. In 1937 and 1938 the Berlin group's problem, as they saw it, was to explain how one could have triple isomers, which moreover engendered others with "inherited" characteristics so that the chains remained distinct. Possible "molecular" models aside, Meitner, Hahn, and Strassmann confessed that their isomers were "very difficult to harmonize with prevailing representations of the nucleus."[33] While this was puzzling, it did not force them to reconsider all their decay schemes. But there were people elsewhere who were more critical.

At the Radium Institute in Paris, Irène Curie had already moved into the neutron irradiation business in 1935, not long after Fermi's group showed the way. With her collaborators Hans von Halban, Jr. and Peter Preiswerk, she irradiated thorium and found several radioactive products, including some not found before by Fermi's group. For example, the Paris team found a substance with a 3.5-hour half-life. Since it behaved like lanthanum in chemical separations, they presumed that it was an isotope of the element that lies beneath lanthanum in the periodic table, namely actinium. This actinium isotope might fill in a long-standing gap in the knowledge of radio-active substances.

Scientists were familiar with three natural radioactive decay series. In the thorium series all the atomic masses were multiples of four, that is, of the form $4n$. In the other series the masses were all of the form $4n + 2$ or $4n + 3$ respectively. The missing $4n + 1$ series, long sought in natural sub-stances, could be expected to result when a neutron was added to the normal isotope of thorium: $^{232}\text{Th} + n \rightarrow {}^{233}\text{Th}$. Thus Curie, Halban, and Preiswerk announced in late May 1935 that they had discovered the missing series.[34]

Hahn and Meitner, however, had been working on the same problem and had already announced in early May, independently, the discovery of the $4n + 1$ series. The Paris group did not accept this announcement as giving

Hahn and Meitner priority in the discovery, for the Berlin team's procedures, Curie felt, were too incomplete and error-prone to constitute proof. Hahn and Meitner, on the other hand, considered themselves the rightful originators of a discovery which Curie's team had simply confirmed and extended.[35] To Rutherford, Hahn wrote that he and Meitner were "a little angry with Mme Irène Curie for not having cited us properly." With an eye to the uneasy political situation in Berlin (pro-Nazi views were held even within the walls of his institute), Hahn added, "We regret this very much, for a scientific echo has never been as necessary to us as just now."[36]

Hahn and Meitner therefore had little reason to be pleased when they learned in 1937 that Curie had moved into the heart of their own territory, the radiochemistry of irradiated uranium. Curie was taking a fresh look at the whole business. While she admitted that the numerous products would be hard to understand without resorting to isomers, she accepted none of the Berlin team's elaborate decay schemes. Once admit the possibility of isomerism, said Curie, and "you can make a great variety of suppositions about the possible transformations."[37] Worse still, Curie and her collaborator Pavel Savitch reported yet another product of irradiated uranium, and one of a type whose existence Hahn and Meitner had consistently denied.

Curie had done this with a delightfully simple experimental trick. She had ignored all the Berlin group's elaborate chemical work and come up with an entirely different approach. Like Hahn and Meitner, she recognized that the most severe problem in investigating irradiated uranium was the buildup of the natural decay products, which soon swamped the radioactivity of the minute quantities of newly produced elements. Hahn and Meitner, we recall, got around this by separating their supposed transuranium elements by precipitation with hydrogen sulfide, leaving behind radium and all its neighboring elements; they had never tried to investigate the increasingly radioactive filtrate. Curie left all the products together in solution as a complex radioactive broth and took up the problem by another handle.

The beta particles emitted by radium and protactinium are relatively weak, with energies around 1 million electron volts or less. They could therefore be blocked with a thin sheet of copper. Curie covered her radioactive broth with thicker and thicker copper sheets until only one sort of radiation got through, and then she studied this strong radiation. It had a half-life of 3.5 hours and was associated with a substance whose chemical properties resembled thorium's.

It did not occur to Curie and Savitch to identify this with the 3.5-hour substance that Curie, Halban, and Preiswerk had two years earlier produced from thorium and called an actinium isotope. For actinium is three places from uranium in the periodic table; an alpha particle emission would move a nucleus only two places, and no known process could produce a greater change in a nucleus. Curie and Savitch thought it likely that they had produced a process of the familiar type in which neutron irradiation induces alpha-particle emission. Meitner and Hahn, as we have seen, had looked for this process and at one point had even thought they found it, but soon eliminated it from their scheme. Now Curie and Savitch said that they had discovered this missing process, and indeed that it produced a substance with strong activity — easy to find if you knew how to look for it.

The Berlin group could scarcely imagine that they had missed such an important isotope in their years of work. They dubbed the 3.5-hour substance "Curiosum." Strassmann at last took a close look at the filtrate. Using thorium as a carrier, he failed, as in the past, to find any radioactive thorium as a product of irradiated uranium. Meitner also asked one of the students, Gottfried von Droste, to try another search for the alpha particles that should be emitted when uranium transmuted to thorium; as usual, none were found.[38] In January 1938 Hahn and Meitner wrote Curie a letter, giving her some details of their new experiments and politely suggesting that she had committed a gross error. Perhaps she had found one of the substances they had already discussed in print, one with a 2.5-hour half-life, and imagined it to be a new substance with a noticeably longer half-life. Courteously they suggested that if she would make a public retraction, they would not publish their criticism.[39]

Curie and Savitch replied with a public retraction which was even more disturbing than their original claim. While admitting that the 3.5-hour substance was not thorium, they reaffirmed its existence. And they said that it followed lanthanum in chemical separations. From the chemical point of view, they declared, it might be either actinium or some new and peculiar element. But "from the physical point of view, both hypotheses run up against considerable difficulties."[40] There was something very odd going on in these uranium products.

In May 1938 Hahn met Curie's husband, Frédéric Joliot, for the first time, at the Tenth International Congress of Chemistry in Rome. As Joliot recalled it, Hahn expressed respect and friendship, but did not hide his skepticism about Curie's work. He told Joliot that he was going to repeat her experiments and prove that she was in error.[41]

Meanwhile Curie and Savitch undertook to study the chemistry of their 3.5-hour substance more closely. Now they performed one of the most

important of all the experiments that led to the discovery of fission. They
had used lanthanum as a carrier. To separate their suspected actinium from
this carrier, they undertook a laborious fractional crystallization: the classic
process (devised years before by Irène Curie's mother, Marie) for separating
the two similar elements. Once the fractions were separated, Curie and
Savitch could readily spot the 3.5-hour substance, watching its radioactivity
decay over the first few hours. Extrapolating back, they could measure how
much of the 3.5-hour substance was present at the outset. After a while,
though, the radioactivity would be expected to reach a minimum and then
begin to increase due to the buildup of natural decay products of actinium.
The strength of this radioactivity after several days was a measure of the
amount of actinium that had been present in the fraction at the outset.

But Curie and Savitch found that the amount of 3.5-hour substance was
not, after all, in proportion to the amount of actinium. Rather, the 3.5-hour
activity tended to appear in the same fraction as the lanthanum. Curie and
Savitch therefore presumed that they had, not actinium, but some new and
singular transuranium element.[42] They failed to imagine that the substance
could be lanthanum itself. Not only was it unthinkable that a uranium atom
could change into an atom half its weight, but also there were signs that
the 3.5-hour substance did not follow lanthanum completely. We can look
back and see that they were measuring a mixture of lanthanum and yttrium,
two rare-earth elements with deceptively similar chemical and radioactive
properties; small wonder that they were confused. By one account, Curie
told a visitor that she felt as if she had all the chemical elements in her broth.
In the laboratories and seminars of Paris, Curie's work was vigorously discussed, without result.[43]

There seemed to be no place in the periodic table, as extended by the
Berlin group, for the 3.5-hour substance. Curie and Savitch had to resort
to hypotheses that they admitted were implausible. For example, they
wondered whether the electronic structure of some of the transuranium
elements might be so delicately balanced that the state of excitation of
a nuclear isomer could affect the distribution of valence electrons and thus
determine the chemical properties. According to their later recollection,
they even entertained the fantasy that the uranium nucleus might somehow
break apart. They rejected this because their new substance was generally
similar to the transuranium elements whose existence the Berlin group
seemed to have proved beyond question.[44]

Other people had as much trouble as the French in understanding the
new data. At the beginning, the work on uranium and thorium in Rome and

Berlin had attracted no more scientific interest than had superficially similar work on other elements — a few lines in review articles and textbooks. Gradually the matter drew attention. Scientists noticed the difficulty of the multiple isomers and the singularity of the transuranium decay schemes with their multiple beta decays. As early as March 1936, Szilard wrote Bohr that he found it difficult to assume so many isomers. Other scientists began to enter the field. A group at the University of Michigan bombarded uranium with neutrons generated by a cyclotron, finding the 3.5-hour decay product and others. They also extended some of the thorium results, as did a pair of scientists at the Vienna Radium Institute. The Vienna scientists expected to find alpha particles emitted after neutron irradiation, but searched in vain.[45] In Berkeley, Ernest Lawrence set Philip Abelson to searching for alpha particles, with the usual lack of success. Abelson recalled that the unorthodox decay schemes for the transuranium elements, with their multiple beta decays so unlike the decay patterns for known heavy elements, "aroused puzzlement and skepticism" in Berkeley. In 1938 Abelson began to search for the X rays emitted by the supposed transuranium elements, since X-ray lines had long been recognized as an unequivocal way to identify an element. Similarly, at Cambridge University Norman Feather and Egon Bretscher prepared apparatus to search for the X rays of the transuranium elements in order to pin down the elements' nature. Interviewed long after, Feather recalled having felt that there was something "very odd" about the transuranium decay schemes. Only a very few isomers were known among the ordinary elements, but beyond uranium a great many appeared all at once. In both Berkeley and Cambridge the experiments began to turn up results that seemed hard to explain.[46] Yet the published literature as of 1938 still showed only confirmations of the work in Paris and Berlin.

According to Strassmann's recollection, Meitner threw up her hands at the Paris results once Curie and Savitch reported that their 3.5-hour substance resembled actinium. A neutron might be able to provoke a nucleus to emit an alpha particle (since these heavy nuclei were almost able to do this spontaneously anyway), but it seemed altogether unlikely that a slow neutron could provoke the emission of an alpha particle and a proton together, which would be necessary to make the jump from uranium to actinium. The Berlin group simply proceeded with their program of chemical studies of irradiated elements. Since both the Paris and Vienna groups had reported new results from thorium, Meitner, Hahn, and Strassmann returned to this element. Always careful and thorough workers, they could see from the problems with uranium that they must be more careful still, checking for every possible element. And indeed they found elements they had not seen before, a whole set of beta-decaying products.

The Berlin group identified the new products as falling in three chains, just like the decay chains starting with uranium, but the new ones apparently starting with three isomers of radium. Presumably each radium nucleus was produced from thorium by the emission of an alpha particle (undetected, however). The table of substances was now more uncomfortably crowded than ever. And the chemical separations, the scientists remarked, were "not altogether easy" and were sometimes unreproducible. The triple isomers still had no explanation in von Weizsäcker's theory. The Berlin team must have seemed, not only to their increasingly skeptical colleagues abroad, but sometimes even to themselves, to be wandering deeper into a maze. Yet they could still hope that their isomers would stand up and even lead to some significant discovery about nuclear properties.[47]

This was the last work Meitner did in the institute. After Austria was joined to Germany in early 1938, Meitner's Austrian citizenship no longer protected her from the persecutions visited on people of Jewish ancestry. Surreptitiously she fled the country, then settled in Sweden. She could keep in very close touch with Hahn by mail, so she remained a real member of the team.

In the Fall, Hahn received the paper of Curie and Savitch announcing that the 3.5-hour substance resembled nothing so much as lanthanum. His reaction, a visitor recalled, was that "it just could not be, and that Curie and Savitch were very muddled up." Strassmann and Hahn now undertook a serious hunt in the products of irradiated uranium, looking for any elements like thorium, actinium, and radium. They found a number of radioactive substances they had hitherto overlooked. In November 1938 Hahn and Strassmann published an account of three new isomers of radium. Each was the start of the familiar sort of beta-decay chain, giving rise to an isomer of actinium. By now, they remarked, they had demonstrated the existence and determined the properties of sixteen different types of nuclei all produced by bombarding uranium with neutrons.[48]

Since several of these products had half-lives in the range of 1 to 4 hours, Hahn and Strassmann concluded that the 3.5-hour substance seen in Paris was in fact a mixture of several isotopes. This would explain the fact that it did not behave quite like any known element. But the apparent solution of the experimental conundrum threw them into worse theoretical difficulty than ever. Above all, how could one explain the creation of radium out of uranium, an element four units higher in atomic number?

Strassmann and Hahn suggested that uranium, on being struck by a neutron, emitted two alpha particles at almost the same time. This was extraordinary enough; to make it worse, the process could be initiated with slow neutrons. Theorists in Berlin pointed out to Hahn that it was barely possible that a slow neutron might initiate a single alpha emission, but there was no way a slow neutron could bring in the many million electron volts of energy needed to get two particles out through the barrier.[49]

Around this time Hahn visited Niels Bohr in Copenhagen, and Bohr too told Hahn that he could not believe a double alpha emission was possible. Bohr urged Hahn to check again that the supposed radium atoms were not some sort of transuranium elements. By letter, Meitner too gave warning and urged Hahn and Strassmann to press ahead with more chemical studies, to be sure to eliminate the known isotopes. As Hahn wrote Feather several months later, "Only after several physicists had expressed their astonishment that slow neutrons should initiate two successive alpha-processes in uranium, did Strassmann and I, in order to dispel the doubts of the physicists, investigate still more carefully the properties of our radium-isotopes."[50]

Through November and the first weeks of December 1938 Hahn and Strassmann studied their mixture of substances, using barium as a carrier for the supposed radium, lanthanum for the actinium. Along the way they found what they took to be yet a fourth pair of radium and actinium isomers. If triple isomers were possible, there was no reason to exclude quadruple ones. But in fact the whole rickety structure was ready to collapse.

Now, Hahn recalled, "since we were dealing with very weak preparations, we undertook to separate the radium isotopes from the carrier substance, barium, to obtain thinner layers of material so that their radioactivity could be more easily measured."[51] Over the next several weeks, the experiments Hahn and Strassmann undertook to study their radium more closely were very similar to the experiments that Curie and Savitch had used to study their 3.5-hour substance, except that Hahn and Strassmann pushed farther. As in Paris, the Berlin separations were to be done by fractional crystallization; as in Paris, the Berlin results were contrary to plan. The method, long familiar to radiochemists for extracting radium, should have concentrated the radiation from the new substances along with the radium, but this failed to happen. Hahn and Strassmann began further chemical tests (for more information on all this period, see the chapter by Fritz Krafft in this volume). Hahn began to call his new substance "the radium isotope with the properties of barium." The distinction between actinium and lanthanum, too, turned out to be as elusive for Hahn and Strassmann as it had been for the Paris team.[52]

The conclusive test, carried out after much groping, began on December 16, 1938. Hahn and Strassmann irradiated uranium and then performed a painstaking fractional crystallization yielding three different fractions, each containing the same amount of the carrier, barium. Since radium crystallizes out at a slightly different rate than barium, they expected to find more of its radioactivity in the first fraction than in the third. In particular, they looked for the 86-minute decay that characterized the substance they called radium III. By the evening of December 17 they had found this activity to have about the same strength in each of the three fractions: it followed the barium. This could simply have indicated that the fractionation was not working properly, but they had undertaken a control, much like the one Curie and Savitch had used when they showed that their 3.5-hour substance differed from actinium. Watching the radioactive curves develop, Hahn and Strassmann could see by noon on December 18 that the fractional crystallization was working properly. Their supposed radium was behaving like no other radium, but like barium.[53]

It is interesting to see how Hahn recorded this result in his notebook, as seen in Figure 13. He apparently intended to write that "the activity of

Fig. 13. Hahn's description of the outcome of the experiment of December 17, 1938: "Die Aktivität von Ba III 86 Min ist auf die erste Abscheidungsz[eit] bezogen für alle!" Inset shows how Hahn wrote "Ra III" and "Ba" on the same page (Hahn-Strassmann notebook, Deutsches Museum, Munich).

radium III is found for all [fractions]!" However, he wrote not "Ra III" but "Ba III" (the difference between Hahn's handwriting for the two elements is small but distinct, as we see elsewhere on the same page). Now, earlier on this same page, and farther ahead in the notebook, he calls this substance "Ra III,' and so it seems he did not deliberately write "Ba III" at this point. Probably his pen slipped.[54]

Slips of the pen are nothing to shrug off. They can provide a rare window into the subconscious, a means by which one can express ideas that the conscious mind resists.[55] In the back of Hahn's mind the substance in question was transforming from a weird sort of radium into simple barium − yet he resisted the thought.

There is adequate evidence of Hahn's resistance in a letter he wrote Lise Meitner on the evening of December 19. By that time continued measurements of the mixture of radium and barium were unequivocal. Hahn was up late, making similar measurements on a mixture of actinium and lanthanum, for here too he was beginning to guess that he had been looking at the wrong part of the Periodic Table. There could be some unlucky coincidence, he told Meitner, but "we come more and more to the frightful [*schrecklich*] conclusion: our radium isotope behaves not like radium but like barium." What was frightful about this was that it would throw three years of work into the wastebasket − all those sixteen substances whose properties had supposedly been determined. In this paragraph Hahn equivocated by first writing "radium isotope" in quotation marks and at the next point writing it without them, then giving the word "radium" in quotes and yet again not. "Perhaps," Hahn continued, "you can propose some kind of fantastic explanation. We ourselves know that [uranium] *cannot* really burst apart into barium."[56] Later on, Hahn clearly recalled his resistance. In 1939 he described his discovery thus: "Against our will, as a result of systematic indicator experiments, we were compelled to conclude that our substance was barium." On later occasions he continued to use this term: "compelled."[57]

Strassmann, on the other hand, did not recall having felt such resistance, although he had earlier experienced it from outside. In 1936 he had done some uranium experiments on his own and found what seemed to be barium. When he told Meitner, she thought barium was altogether unlikely and told him he might as well throw away his notes. Thus, Strassmann recalled, when the barium appeared again in December 1938 he was ready to accept it.[58]

It was Hahn who wrote up the results, and he avoided a definite statement. "As chemists," he wrote, he and Strassmann should simply take their whole scheme and replace the symbols for radium, actinium, etc. with the symbols for barium, lanthanum, etc. But "As 'nuclear chemists' closely associated in a certain measure with physics, we cannot yet commit ourselves to taking this leap, which is contrary to all previous laws of nuclear physics." On December

22 Hahn took the manuscript of this paper to his friend Paul Roshaud, editor of *Die Naturwissenschaften*, who agreed to hurry the paper into print.[59]

A few days later a further thought occurred to Hahn. "Would it be possible," he wrote Meitner, "that uranium 239 bursts into a barium and a masurium." Masurium (i.e., technetium) was the element whose atomic weight would be given by the difference between the weights of barium and uranium; chemically it might well have been masquerading all along as their eka-rhenium.[60] This thought, that uranium might split in two, may have already occurred to Meitner and her nephew Otto Frisch, a physicist who was taking his Christmas holidays with his aunt and was puzzling with her over Hahn's letters. Using the liquid-drop model of the nucleus, they could start to understand theoretically what was happening. On the first day of 1939 Meitner replied to Hahn that although she had strong reservations, "it *may* be energetically possible that such a heavy nucleus could burst apart."[64]

At about this time Hahn put his idea into print by making changes in the proofs of his and Strassmann's article in *Die Naturwissenschaften*. Hahn added in proof the remark, "The sum of the mass numbers Ba + Ma, thus e.g. 138 + 101, gives 239!" And he changed his comment that the experimental results were "contrary to all previous laws of nuclear physics" to read, "contrary to all previous experience in nuclear physics."[62] The laws had fallen.

After Hahn and Strassmann's paper was published, nuclear fission was rapidly accepted. Frisch, returning to Copenhagen, began explaining fission to Bohr, when Bohr interrupted, striking his head with his fist and exclaiming, "Oh, what idiots we have been that we haven't seen that before." In Paris, Joliot took a few days to puzzle over the announcement that barium was present, then explained the results to Curie. "What fools we were!" she exclaimed. In Berkeley, Abelson's thesis advisor saw a newspaper report on fission and ran to tell Abelson that he had barely missed a big discovery; Abelson went into a daze. While one or two scientists resisted believing in fission until they saw experimental confirmation (which only took a few days), most physicists accepted the explanation as soon as it was offered.[63]

For the discovery of fission fitted neatly into physicists' developing ideas about the liquid-drop model of the nucleus and tended to solidify these ideas. Scientifically speaking, in light of present knowledge fission is simply a peculiar instability of a few heavy nuclei — just one more nuclear reaction in the list that was developed in the 1930s, and not the most theoretically fundamental reaction either. It is worth asking, then, why scientists took such a long time to recognize fission, and why they were all so surprised when it was revealed. The history may be worth another look, with an eye to the general question of how scientists do their work. From this modest case history we may be able to learn as much about certain important modes of scientific work as we would learn from the study of one of those immense, complex developments that transforms an entire science.

I suggested at the start that scientific work usually resembles puzzle solving. Certainly in the history of fission we see a string of puzzles confronted. By a puzzle I mean a problem of restricted scope, stated clearly enough so that we may check that the solution does satisfy the conditions set in the problem. The solution may add an increment to knowledge or even overturn something previously thought to be true, but the solution will not change the way we define the puzzle. Then the question is, what are the rules for defining such puzzles and recognizing the solutions? The set of concepts that governs the approach to a particular range of puzzles may be called a *paradigm*.

Here I must pause to acknowledge that I am drawing on ideas made notorious by Thomas Kuhn in his *Structure of Scientific Revolutions*. Kuhn describes something he calls "normal" science, which he calls a "puzzle-solving" activity. These puzzles are both defined and solved by what Kuhn calls a "paradigm."[64] Kuhn's ideas have had a curious career: refuted thoroughly time and again, they keep returning to life. Scientists in particular find something irresistibly plausible in his description of science. Let us admit, then, that Kuhn may say something worth listening to, even if we do not agree with every part of his description of how science develops.

I will steer clear of these debates so far as possible, and in particular I will avoid using the term "normal science," a term some find condescending. Hahn, Meitner, Curie, and the rest were all doing normal science by Kuhn's definition, yet their work was no less extraordinary and important to the progress of nuclear physics. When I say that they were engaged in solving puzzles, I am not implying that there was anything trivial about their work. I am simply drawing attention to the set of concepts, the paradigm, that governs the approach to a scientific problem much as rules govern an everyday puzzle. Kuhn did not invent the idea nor even the term for a paradigm, and others (such as Imre Lakatos with his "research programmes") have developed related ideas. In following Kuhn I am not claiming that his approach is uniquely good, but I am simply pointing out ways it may be useful. I am indebted to Kuhn for refining the old idea of paradigms, for bringing it forcefully to our attention, and for giving hints as to where actual paradigms may be discovered.[65]

A paradigm, as Kuhn emphasizes, is often not spelled out completely but is learned implicitly in the course of studying how previous puzzles were defined and solved. Therefore we may expect to find paradigms most clearly stated where instruction is taking place. For a paradigm is nothing if it is not a widely held set of concepts that can be communicated to people entering the field. Thus I propose to look in textbooks, introductory lectures, review articles, and similar materials for instructing novices. Historians of science rarely give full attention to such materials, even though we may find that review articles and lectures, sometimes even textbooks, can be as up-to-date as the journal literature proper.

In the 1930s, textbooks and the like usually gave much attention from the start to experimental results. Many of the experiments, and nearly all those we are concerned with here, followed a simple pattern. One bombarded a substance with particles and studied what came out. In many cases there was a transmutation within the substance as well as emission. This sort of bombardment had produced the discovery of the neutron, artificial radioactivity, and a host of other results. There were various puzzles set by the procedure, but the one that concerns us stemmed from the fact of transmutation. This set the puzzle of determing what the product was, its atomic number, and perhaps other chemical and physical properties. The problem was simplified by the confident expectation that the product would be found among the restricted set of elements whose atomic numbers were within a few units of the atomic number of the bombarded substance. This was exactly the sort of task that chiefly concerned not only the Berlin group but the Rome and Paris groups and others too.

Note that the experimental procedures used in the work leading to the discovery of fission had little to do with the cyclotrons and other advanced tools of the thirties. Even the final, crucial experiments by Hahn and Strassmann relied chiefly on standard techniques. Hahn had used much the same techniques since before World War I; Strassmann, Irène Curie, and others learned them as part of their initiation into radiochemistry. This is not to say that the experiments were easy. To detect and identify a few thousand atoms, using the balky homemade counters of the period, required consummate skill — but little new technique.

On the theoretical side, the latest models and equations did not, as we have seen, hasten the discovery of fission any more than did the latest experimental gadgets. The concepts that governed the studies of uranium were well tried, and to this day people who know their nuclear physics will find these concepts familiar, even obvious. They are readily found in textbooks and other instructional materials.

Figure 14 shows an illustration in Gamow's popular elementary textbook

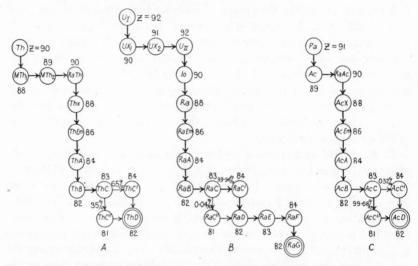

Fig. 14. Radioactive decay chains portrayed as moves on a checkerboard; atomic number increases upward and atomic mass to the right. (George Gamow, *Constitution of Atomic Nuclei and Radioactivity*, Oxford: Clarendon, 1931, p. 30.)

of 1930. In this diagram the transmutations of one element into another in natural radioactive decay are portrayed like moves on a checkerboard, horizontal for changes in atomic mass and vertical for changes in atomic number. Note that in this scheme it is the transmutation that counts; the beta or alpha particle emitted in the process is not shown but only implied by the direction of the arrow. When artificial transmutations joined the naturally occurring ones, diagrams like Gamow's did not have to be changed fundamentally; one simply added new arrows, as in the illustration from von Weizsäcker's textbook shown in Figure 15. Besides alpha and beta emission von Weizsäcker shows a number of other possible processes, all of which involve a step of no more than one or two units. Another example of this sort of elaborate scheme is shown in Figure 16, from the "orange bible" review articles of Hans Bethe and his collaborators, the most used of all introductory material on nuclear physics in the latter thirties. (Here mass number increases right to left, the inverse of Gamow's convention.) Note off to one side the little charts showing the types of transformations known to exist, represented as short arrows in various directions.

There is simply no way to fit nuclear fission conveniently into such a scheme. In the first place we would have, not short jumps of one or two

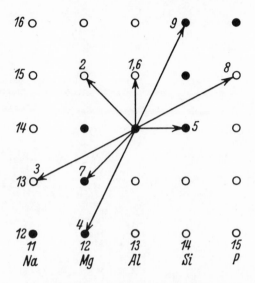

Fig. 15. The known transformations of $^{27}_{13}$Al to nearby isotopes. (Carl Friedrich von Weizsäcker, *Die Atomkerne: Grundlagen und Anwendungen ihrer Theorie*, Leipzig: Akademische Verlagsgesellschaft, 1937, p. 103)

units, but enormous leaps half the length of the table. In the second place the leap would have not one but two endpoints, the two fragments resulting when the nucleus splits apart. Here we see the consequences of failing to represent fully the particle emitted in the course of transmutation. To the extent that scientists of the 1930s held a diagram of this type in their heads as they worked, they would have had difficulty in picturing fission.

Hahn and his collaborators did know how to use such diagrams,[66] but the Berlin group and many others tended to use a different scheme to represent transmutation. Their notation resembled the checkerboard diagram collapsed into one dimension. Figure 17 shows how it was presented in the classic text of Rutherford, Chadwick, and Ellis. The particle emitted during a transmutation is not entirely neglected here, but it takes a back seat. A more complete notation came to the fore after artificial radioactivity reactions burgeoned. Drawn from the familiar way of representing chemical reactions, it was what Hahn called the "chemists' notation."[67] I take as an example Figure 18, from Marie Curie's definitive text, edited by Irène Curie after her mother's death. The equation shows the artificial transmutation of aluminum when bombarded with alpha particles (helium), resulting in a silicon isotope, a

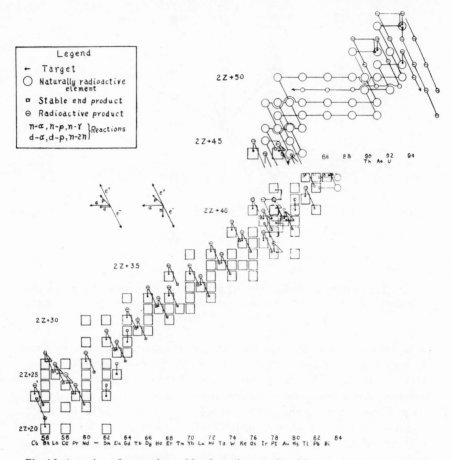

Fig. 16. A section of a complete table of reactions produced by neutrons and deuterons, as known in 1937. (M. S. Livingston and H. Bethe, *Rev. Mod. Phys.,* 9 (1937), 294).

$$\text{Radium A} \xrightarrow{a} \text{Radium B} \xrightarrow{\beta + \gamma} \text{Radium C} \xrightarrow{a + \beta + \gamma}$$
$$3 \cdot 05 \text{ min.} \qquad 26 \cdot 8 \text{ min.} \qquad 19 \cdot 7 \text{ min.}$$

Fig. 17. Rutherford's linear notation for showing radioactive decay. (E. Rutherford, J. Chadwick, and C. D. Ellis, *Radiations from Radioactive Substances,* Cambridge: Cambridge University Press, 1930, p. 20.)

$$^{27}_{13}Al + ^{4}_{2}He = ^{30}_{14}Si + ^{1}_{1}H + \Delta$$

Fig. 18. "Chemical" notation for bombardment of aluminum with an alpha particle (helium). (M. Curie, *Radioactivité*, Paris: Hermann, 1935, p. 373.)

proton (hydrogen), and an amount Δ of energy. Incidentally, this was highly paradigmatic: it was the reaction used by the Joliot-Curies in their discovery of artificial radioactivity, and in the latter thirties it was the most common textbook example of how this notation worked. Teachers pointed out that this sort of equation is beyond reproach, being derived in principle from the conservation of mass-energy. For variety I show in Figure 19 an example drawn not from a textbook but from lectures that Meitner gave at about the same time.[68]

$$m_A + m_\alpha + E_\alpha = m_B + m_H + E_B + E_H + Q. \qquad (8)$$

Fig. 19. Fundamental conservation equation underlying notations for nuclear transformations (*m* for mass, *E* for energy of particle, *Q* for nuclear energy loss or gain). (Lise Meitner and Max Delbrück, *Der Aufbau der Atomkerne*, Berlin: Springer, 1935, p. 14.)

In principle, this general equation can easily accommodate fission. But the workers in the 1930s could scarcely get far with a general conservation equation; for definiteness they used specific cases. Each case covered one type of reaction. In all of nuclear physics, something like eighteen types of reactions were known,[69] but for neutron bombardment alone very few results were known or even suspected to exist. Figure 20 shows how Fermi listed the

$$^{M}_{Z}A + ^{1}_{0}n = ^{M-3}_{Z-2}A + ^{4}_{2}He$$
$$^{M}_{Z}A + ^{1}_{0}n = ^{M}_{Z-1}A + ^{1}_{1}H$$
$$^{M}_{Z}A + ^{1}_{0}n = ^{M+1}_{Z}A$$

Fig. 20. Generalized description of reactions of a neutron with a nucleus of mass number *M* and charge *Z*. (E. Fermi (1938) in *Nobel Lectures: Physics, 1922–1941*. (Amsterdam: Elsevier, 1965, p. 416.)

principal neutron reactions in his 1938 Nobel lecture. Here are nuclei in the abstract, reduced to mass numbers M and Z. Bombardment by neutrons gives three possible results: alpha-particle emission, proton emission, or resonant capture of the neutron. This sort of listing was not unique to Fermi. On page 98 we saw how Hahn and Meitner listed such reactions, using aluminum for their example, and in Figure 21 we see how Meitner wrote the list in the abstract, in the same order.[70] Meitner has been more complete than Fermi, showing the expected emission of electrons (beta particles).

$$
\begin{aligned}
&\text{I.}\quad A^z + n \rightarrow (A - 3)^{z-2} + \alpha \\
&\qquad\qquad\qquad\searrow \\
&\qquad\qquad\qquad (A - 3)^{z-1} + e^- \\
&\text{2.}\quad A^z + n \rightarrow A^{z-1} + \mathrm{H}_1^1 \\
&\qquad\qquad\quad\searrow \\
&\qquad\qquad\quad A^z + e^- \\
&\text{3.}\quad A^z + n \rightarrow (A + 1)^z \rightarrow (A + 1)^{z+1} + e^-
\end{aligned}
\qquad (\mathrm{II})
$$

Fig. 21. Another generalized description of neutron reactions, showing the expected radioactive beta decay of the products (emission of e^-). (Meitner and Delbrück, *Aufbau der Atomkerne* (1935), p. 20.)

Evidently this set of equations specifies a research program. At the start one knows the value of atomic mass A and atomic number Z of the bombarded substance; for uranium, $Z = 92$ and $A = 238$ or conceivably 235. One then inspects the results of neutron bombardment and fits the products into one or another of the three types of reactions. The repeated searches in the latter thirties for alpha particles, for example, were searches for behavior of the first type.

This is the way the Berlin group behaved. As we have seen, they came to believe that only the third type of reaction, neutron capture, could accommodate all their findings. Figure 12 is an example of how they wrote their results. The emitted particle has again shrunk to a simple beta above an arrow; it would be no obvious matter to alter such an equation in order to show fission.

Of course, no one would claim that the history of fission can be understood in terms of nothing more than these elementary diagrams and equations. A great deal of theoretical thought lay behind scientists' understanding of alpha-particle emission, beta decay, resonant capture of neutrons, and so forth. And notation was connected not only to theory but to a large body of experimental procedures which gave meaning to each of the terms. One reason I have shown two variant ways of writing transmutations, as jumps on a checkerboard and as equations, is to show that more than simple notation was involved. To make this clearer, let us inspect a third form of notation that was occasionally used. The equation of Figure 18, for example, might also be written: $^{27}_{13}\mathrm{Al}(\alpha, p)^{30}_{14}\mathrm{Si}$, where the parenthetical expression is to be read "alpha particle in, proton out." The other known reactions could be written similarly. Thus Figure 22, from von Weizsäcker's textbook, lists the (n, α). Energetisch günstig bei geradem Z und ungeradem N, aber auch bei einer großen Zahl anderer Elemente gefunden.

(n, d). Nicht beobachtet. Vgl. (α, d).

(n, p). Wird nur von Neutronen großer Primärenergie ausgelöst. Die Erklärung liegt darin, daß er — analog zum Prozeß (p, n) — durch β⁺-Zerfall spontan stattfinden wird, wenn er überhaupt exotherm ist. Dementsprechend sind die Folgeprodukte, die durch ihn aus stabilen Kernen hergestellt werden, ohne Ausnahme β⁻-labil.

(n, γ). Bei allen Kernen energetisch sehr günstig.

Fig. 22. List of conceivable neutron reactions in the (n, x) notation; that is, neutron in, x out. (Von Weizsäcker, *Atomkerne* (1937), p. 105.)

reactions resulting from neutron bombardment. Incidentally, von Weizsäcker clearly felt that this list exhausted the possibilities: he included the neutron-in, deuteron-out (n, d) reaction for completeness, even though it was "not observed." In this notation, even more than in the other two, there is no straightforward way to portray fission.

These notations could all be used almost interchangeably, and they shared key features such as the difficulty in expressing the sort of gross changes that fission brings. Since the notations had much in common, we can infer that they all corresponded to some deeper concept of nuclear processes. This concept is what we might name *transmutation*: an emphasis on a change in the nucleus. Outgoing particles are significant because they manifest the change in the nucleus; incoming particles can provoke the change, but it may happen spontaneously, as in the naturally radioactive elements. People assumed that the transmutation was a relatively minor change, leaving the nucleus only a little different from the way it was before.[71]

I have attempted to abstract the transmutation concept in a general way, but no doubt many physicists had a concrete mental picture of the nucleus. Perhaps they saw a nucleus as something like a bag or a dish full of marbles (see the Chapter by Roger Stuewer in this volume). A particle thrown into the dish might knock a particle out, but this would not much change the overall condition of the nucleus. However, the scientists who figure in the fission story generally stuck to formal notation and did not draw pictures, so we cannot say whether concrete representations played much of a role in their thinking.

In sum, Hahn, Meitner, and many of their colleagues worked with a paradigm, which had several components. There was a concept of transmutation, possibly connected with some sort of mental picture of nuclei. There was a choice of notations by which transmutations could be represented. And there were cases showing how to use a notation, including quite specific cases like the example of artificial radioactivity in silicon (Fig. 18) and cases expressed in quite general terms of nuclei characterized only by A and Z (Fig. 21). All these things, and possibly more that are harder to pin down, comprised the paradigm used in work on uranium.[72]

The paradigm was not used rigidly but was modified to agree with new observations. Almost from the start, Meitner and Hahn were willing to postulate reactions slightly different from any reaction seen before. In their early 1936 paper, as we have seen, they listed the three known types of neutron reactions but showed that not all their results could fit into this pattern. So they postulated a fourth type of reaction, not so different from the three already known, wherein two neutrons come flying out together. This reaction turned out to be a valid one, another case in the collection of cases that was part of the paradigm. When the double neutron reaction nevertheless failed to explain their results, the Berlin group turned to the concept of isomers. Meitner took one of the accepted types of reactions and extended it, used it as a template so to speak, to accommodate not one nucleus but two. Since isomerism does exist, the expansion of the paradigm was again valid, a permanent addition to knowledge of the nucleus.

This sort of work on a paradigm has been discussed briefly by Kuhn. He points out that as a scientific field is explored, anomalies may appear, observations that do not exactly fit the paradigm. Scientists explore the area of the anomaly and "adjust" or "articulate" their paradigm. It may also happen, as Kuhn notes, that resistant anomalies appear, puzzles so difficult to solve with the apparatus of the paradigm that scientists become uneasy.[73] This description is accurate for the history of fission. We have seen that by 1938 uranium was a serious headache, with the spread of misgivings evidenced by the entry of Curie, Abelson, Feather, and others into the field. New explanations were proposed, each more distant than the last from previous cases of nuclear behavior. Triple and then quadruple isomerism; a jump of three atomic numbers; a connection between isomeric states and chemical properties; a jump of four atomic numbers by double alpha-particle emission; at last, fission — these were the changes in the paradigm, each hypothesis wilder than the last, which scientists proposed in attempts to resolve the uranium puzzle.

Eventually Hahn harked back to the form of the equations that represented simple conservation of mass. (Perhaps his mental picture of nuclei, as a bag of particles or whatever, was also generalized.) His problem was solved once he could see the right-hand side of the equation as barium plus another medium-weight element. The importance of this basic form to Hahn is shown by the fact that his first impulse was to use this equation for mass rather than an equation for atomic number, even though, as he soon realized, it is the atomic number equation that should be used to determine the chemical nature of the fission products.

It appears that at the point when Hahn put barium into the equation the paradigm served as a handy framework for his thought but did not dominate his viewpoint. Rather, the experimental results took over. Barium thrust itself into the equation and reshaped it, "compelled" the equation to take a shape closer to reality. If we try to see just how this happened, we find ourselves on the threshold of Hahn's subconscious thinking. There we must halt. The search for a moment when fission was "discovered," a magical gestalt shift, is a romantic but futile quest; we can only see a gradually strengthening, unwelcome suspicion that things were not as everyone had supposed.

In the end the paradigm was not altogether overthrown. Even the notations, so inadequate to portray fission, remain in use to this day for other types of nuclear processes. Superficially, fission simply expanded the paradigm to include another case, another nuclear process in the list.

But the new case was different from the rest. The tiny recognition that barium was present in a dish of chemicals had great leverage, setting off a chain of recognitions that forced scientists to realize that the nucleus is capable of greater convulsions than they had previously imagined. This gave the liquid-drop model not only its first big success but also a new depth of meaning. The exasperated surprise of Bohr, Curie, Abelson, and many others when they learned the solution to the uranium puzzle is the surprise of people who realize that an old assumption is false. The assumption that nuclear reactions are always transmutations — that is, changes that leave the bulk of a nucleus practically untouched — had to be abandoned, and replacing this notion entailed the utter collapse of the elaborate Berlin scheme of transuranics and isomers. Is this an example of a scientific revolution?

Kuhn originally described scientific revolutions as events in which one paradigm is abandoned in favor of an entirely different paradigm, in the teeth of resistance by an old guard. Such a complete supersession of paradigms and such a struggle never happened in the entire history of nuclear physics. In the matter of fission, most nuclear reactions were still understood the same way as before, and the only resistance that had to be fought was the resistance that Hahn and his colleagues had to overcome before the discovery could be made in the first place.

Now, Kuhn himself accepts that scientific advance need not always involve struggle over a wholesale replacement of ideas. He would probably argue that the fission history exemplifies a more limited but still significant revolution, in which parts of a paradigm survive while other parts are everthrown and replaced. Kuhn's critics would reply that this argument dilutes the concept of revolution beyond recognition, so that one could call any scientific change a series of infinitesimal revolutions. I beg leave to avoid commenting on this question of revolutions. In the particular case of fission, I will only point out that there does appear to have been a discontinuous change involving part, but by no means all, of a paradigm.

Even setting aside the question of supposed revolutions, I am not suggesting that Kuhn's description of science is universally valid. Other descriptions may fit the fission history equally well, and other scientific histories may be found to develop in quite different ways.[74] What I do suggest is that the concept of a paradigm, as developed in part by Kuhn, can give us insight into some stories of scientific puzzle solving.

There is a deeper reason to believe that paradigms can be useful. The term "paradigm" comes from an analogy to the paradigms of language study, such as the *amo, amas, amat* pattern that we memorize and then use as a model for other verbs we need to conjugate. The analogy is meaningful. Modern linguistics advises us that all grammar is transformational, that every sentence we utter is formed by a series of substitutions into one or another underlying grammatical pattern. Note the similarity to the paradigm I have displayed for the work leading to the discovery of fission. Here, too, is a pattern, with spaces into which one could fit various atomic numbers, or perhaps various types of reactions: in short, it has a structure somewhat like a grammar. How could it be otherwise? Such schemes are means of communication, whether between teacher and student or between colleagues; therefore they must have characteristics of a language. That explains why paradigms are fit subjects for historical study. Unlike the ineluctable mental representations of some one individual, these are community property, designed for communication. And so they can communicate to us as well.

Having come this far, let us see how much farther we can go by exploiting the paradigm we have identified. I suggest that its interest is not exhausted once we have seen how it sometimes hindered and sometimes helped the process of science. A look at where the paradigm came from can suggest matters beyond the narrow scope of 1930s uranium studies.

It will be recalled that the set of nuclear reactions could be expressed in several different formal ways, and that therefore there had to be a more general concept behind the notations, a concept I have named transmutation.

It is easily traced to its early appearances in nuclear physics, in the work of Ernest Rutherford and Frederick Soddy. Rutherford always saw the nucleus concretely, and we see in Figure 23 how he presented an almost tangible

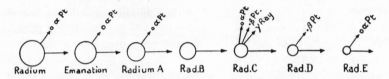

Fig. 23. Primitive form of notation for nuclear transformation (1904). Cf. Figure 17. (published in *Phil. Trans.* A204).

model, in which an alpha particle, for example, appears as a little speck emitted by the transmuting nucleus. It is a short step from this to the equations in which the emitted particle is shown by a small symbol above an arrow (as in Fig. 17). Turned to the vertical, as Rutherford himself showed, this picture is well on its way to becoming a diagram of the decay series in the checkerboard style (like Fig 14).[75] This eminently plausible model of the nucleus, one might almost say this true picture of transmutation, probably underlay every description of nuclear transmutation presented from 1905 on into the 1930s.

It is suggestive that this process was named transmutation. Beginning with Soddy and Rutherford, scientists frankly compared nuclear physics with alchemy, for an essential part of alchemy was the observation that a material substance may change its nature and that this change is often accompanied by a transfer of energy, usually heat. Indeed this observation predates alchemy, since it is obviously related to our familiar experiences with food: the transformation of substance into ourselves with accompanying release of energy. We might expect this complex of ideas, then, to be tied up with the question of bodily health, and indeed from ancient times a chief goal of alchemy was the prolongation of life.[76] Scientific chemistry of the eighteenth and nineteenth centuries abandoned the quest for transmutation of elements and immortality, yet carried forward intact the ancient complex of concepts, for changes in the form of matter (as in metallurgy and other industrial processes) and pharmacology were at the marrow of the discipline of chemistry. As soon as nuclear transmutation was understood — the word was probably first used in this context by the chemist Soddy — its connection with the release of energy was remarked upon. Within a decade, that is, by 1913, the public was fully aware of the peaceful and warlike potential of a vast new store of energy. The medical aspect did not lag, for the notion that radioactivity could be used to prolong life had a durable popularity that went far beyond anything the scientific evidence could warrant.[77] In short, there is reason to suspect that our paradigm has deep roots, that the equation expressing conservation of mass-energy in a transmutation is but one face of a closely linked set of ideas that are historically and psychologically primitive. With this possibility in mind, let us cast one more brief look at the uranium studies of the 1930s.

Consider first the laboratories in which these studies were carried out — large, expensive, well-equipped institutes. The funding and building up of these laboratories were closely connected to hopes for applications of the results of research, not the least of which were hopes for the release of nuclear energy and for the improvement of health. The connections are clearly seen in the histories of the institutions we have discussed in connection with the history of fission — Berkeley, Cambridge, Paris, Rome.[78] But we should give our chief attention to Hahn's laboratory. From the beginning the Berlin Chemistry institute was funded almost entirely by the chemical industry, and without the help of the I. G. Farben company in particular Meitner could never have established her section for radioactive physics. Throughout their work on uranium, Meitner, Hahn, and Strassmann called on industry for help; for example, they were indebted to the Philips laboratory for uranium metal.[79] It would be naive to imagine that industrialists never expected anything in return. It is true that Hahn's institute, like the laboratories in Berkeley, Cambridge, and elsewhere, was supposed to be devoted in the first place to pure science. But this was within the context of a firm belief that pure science (or at least some types of pure science, like nuclear physics) were likely to produce applications. The Berlin institute was largely devoted to subjects that, as Hahn said, "beside purely scientific inquiries also take up an attack on concerns of industrial importance." Until he began to work on uranium, Hahn had been devoted to applied radiochemistry, that is, to the use of radioactive materials for chemical studies with potential practical value, and even while he worked on uranium he continued to supervise students who were doing such work. Radioactive tracers, he felt, would prove of great value not only to pure chemistry and industry but to biology as well. Meitner, too, pointed out that artificial radioelements provided new means for studying chemical and biological processes.[80]

Thus the set of concepts clustering around the idea of transmutation not only provided an intellectual framework for dealing with the uranium puzzle, but also made other contributions to the work. Motivational and material support for the scientists came from various sources, but prominent among these were hopes of producing new chemical products and processes, new benefits for health, and perhaps a new source of energy. Without these hopes it is questionable whether the work leading to fission could ever have been supported. So it is no coincidence that today the chief peaceful uses of nuclear fission are in the generation of energy and the production of radioelements of value to medicine and industry; that is where the work was aiming long before fission was imagined.[81]

It may seem surprising, even implausible, that I find a paradigm so thoroughly embedded in the work that led to nuclear fission. It may feel more comfortable to believe that scientists work in a more random fashion, without much intellectual substructure, still less a structure that is connected with the scientists' motivation and with the money that pays their salaries. If there is no such structure, then they, and by implication we, are exempt from responsibility for considering how our intellectual products may interact with the real world. I assert, however, that paradigms (or whatever else one may name these structures) are there waiting for us to uncover them. We may learn things from them that we might otherwise not understand.

American Institute of Physics

NOTES

I am very grateful to Tania Oster for research assistance and to O. R. Frisch, Lew Kowarski, Fritz Krafft, Thomas Kuhn, Fritz Strassmann, and Roger Stuewer for helpful suggestions. This work was supported in part by the American Institute of Physics.

1 The most complete studies are Hans G. Graetzer and David L. Anderson, eds., *The Discovery of Nuclear Fission* (New York: Van Nostrand Reinhold, 1971); Horst Wohlfarth, *40 Jahre Kernspaltung: Eine Einführung in die Originalliteratur* (Darmstadt: Wissenschaftliche Buchgesellschaft, 1979). See also Ernst H. Berninger, *Otto Hahn in Selbstzeugnissen und Bilddokumenten* (Reinbek bei Hamburg: Rowohlt, 1974); David Irving, *The Virus House: Germany's Atomic Research and Allied Counter-Measures* (London: Allen & Unwin, 1973), pp. 19–30; Esther B. Sparberg, "A Study of the Discovery of Fission," *Amer. J. Phys.*, 32 (1964), 1–8; Louis A. Turner, *Rev. Mod. Phys.* 12 (1940), 1–29. For Kuhn see notes 64, 65 below.

2 L. Meitner, "Einige Erinnerungen an das Kaiser-Wilhelm-Institut für Chemie in Berlin-Dahlem," *Naturwissenschaften*, 41 (1954), 97–99 (p. 98); A. Flammersfeld, "Zur Geschichte der Atomkernisomerie," pp. 74–77 in O. R. Frisch, F. A. Paneth, F. Laves, and P. Rosbaud, eds., *Trends in Atomic Physics: Essays Dedicated to Lise Meitner, Otto Hahn, Max von Laue* . . . (New York: Interscience, 1959), p. 75; Karl-Erik Zimen, "Einige Erinnerungen an das Kaiser-Wilhelm-Institut für Chemie," pp. 79–84, in *ibid.* (p. 83).

128   SPENCER R. WEART

3 Emilio Segrè, *Enrico Fermi, Physicist* (Chicago: University of Chicago Press, 1970), pp. 73–77.
4 A. V. Grosse and M. S. Agruss, *Nature*, **134** (1934), 773. Controversy: *Naturwissenschaften*, **19** (1931), 738, **20** (1932), 362–363, 505–506; also O. Hahn, *Ber. Deut. Chem. Ges.*, **68B** (1935), 478–479. Lise Meitner and Max Delbrück, *Der Aufbau der Atomkerne: Natürliche und künstliche Kernumwandlungen* (Berlin: Springer, 1935), p. 26; Otto Hahn, *My life: The Autobiography of a Scientist*, trans. Ernst Kaiser and Eithne Wilkins (New York: Herder & Herder, 1970) (from *Mein Leben*, Munich: Bruckmann, 1968), p. 148.
5 O. Hahn and L. Meitner, *Naturwissenschaften*, **23** (1935), 37–38, 230–231.
6 I. Noddack, *Angew. Chem.*, **47** (1934), 654.
7 L. L. Quill, "The Transuranium Elements," *Chem. Rev.*, **23** (1938), 87–155, p. 120; Oscar d'Agostino and Emilio Segrè, *Chim. Italia*, **65** (1935), 1096; K. Diebner and E. Grassmann, "Künstliche Radioaktivität," *Phys. Z.*, **37** (1936), 378 [for their other reviews see *ibid.*, **38** (1937), 406–425, **39** (1938), 469–501]. See oral history interview of Segrè by Charles Weiner, 1967, at American Institute of Physics (AIP), pp. 24–25; oral history interview of Grosse by Weiner, 1974, AIP, passim; Walter Gerlach, quoted in Dietrich Hahn, ed., *Otto Hahn: Begründer des Atomzeitalters. Eine Biographie in Bildern und Dokumenten* (Munich: List, 1979), p. 144.
8 O. Hahn and L. Meitner, *Naturwissenschaften*, **23** (1935), 230–231, 544–545; E. Amaldi, O. D'Agostino, B. Pontecorvo, and E. Segrè, *Ricer. Scient.*, **6(1)** (1935), 435–437, in Enrico Fermi, *Collected Papers (Note e memorie)*, E. Segrè, *et al.* (eds.), Vol. I (Chicago: University of Chicago Press, 1962), p. 667.
9 O. Hahn, *Chem. Ber.*, **67A** (1934), 150, 163; O. Hahn, *Applied Radiochemistry* (Ithaca, N.Y.: Cornell University Press, 1936), pp. v–vi. For the Kaiser Wilhelm-Gesellschaft see Brigitte Schroeder-Gudehus, "The Argument for the Self-Government and Public Support of Science in Weimar Germany," *Minerva*, **10** (1972), 537–570; Kaiser Wilhelm-Gesellschaft zur Förderung der Wissenschaften, *Handbuch*, Adolf von Harnack (ed.), (Berlin: Reimar Hobbing, 1928), pp. 31, 40–46; and below, n. 80.
10 L. Meitner, "Wege und Irrwege Zur Kernenergie," *Naturw. Rund.*, **16** (1963), 167–169 (p. 167); *Naturwissenschaften*, **22** (1934), 759, **23** (1935), 37. See also Hahn, *My Life*, p. 147.
11 See also the article by Fritz Krafft in this volume. Rutherford: see below, n. 36. Carl Friedrich von Weizsäcker, *Die Atomkerne: Grundlagen und Anwendungen ihrer Theorie* (Leipzig: Akademische Verlagsgesellschaft, 1937), p. 3; Flammersfeld, "Atomkernisomerie," p. 75.
12 Quill, *Chem. Rev.*, **23** (1938), 100–105; O. Hahn, "Die 'falschen' Transurane. Zur Geschichte eines wissenschaftlichen Irrtums," *Naturw. Rund.*, **15** (1962), 43–47 (p. 44); oral history interview of Grosse by Weiner, 1974, AIP, pp. 40–47.
13 Nov. 12, 1936, Rutherford Correspondence, Cambridge, from microfilm at AIP. Similarly, Meitner and Hahn to Rutherford, Feb. 19, 1937.
14 Oral history interview of Lise Meitner by O. R. Frisch and Thomas Kuhn, 1963, Archive for History of Quantum Physics (copies at AIP; American Philosophical Society, Philadelphia; Niels Bohr Institute, Copenhagen; University of California, Berkeley; and University of Minnesota, Minneapolis).
15 L. Meitner and O. Hahn, *Naturwissenschaften*, **24** (1936), 158–159; O. Hahn, L. Meitner, and F. Strassmann, *Chem. Ber.*, **69B** (1936), 905–919.

[16] E. Fermi, E. Amaldi, O. D'Agostino, F. Rasetti, and E. Segrè, *Proc. Roy. Soc.*, **A146** (1934), 483, in Fermi, *Collected Works*, Vol. I. p. 744. They abandoned this explanation in Amaldi, et al., *Proc. Roy. Soc.*, **A149** (1935), 522–558, in Fermi, *Collected Works*, Vol. I, p. 775. L. I. Rusinov, *Phys. Z. Sowjet.*, **10** (1936), 219–222; F. A. Heyn, *Nature*, **138** (1936), 723; M. S. Livingston and H. Bethe, *Rev. Mod. Phys.*, **9** (1937), 245–390. See also C. H. Johnson and F. T. Hamblin, *Nature*, **138** (1936), 504.

[17] O. Hahn, *Chem. Ber.*, **69A** (1936), 217–227.

[18] G. Gamow, *Constitution of Atomic Nuclei and Radioactivity* (Oxford: Clarendon, 1931), passim; cf. Gamow, *Structure of Atomic Nuclei and Nuclear Transformations* (Oxford: Clarendon, 1937), pp. 16–18. Georg von Hevesy and F. A. Paneth, *A Manual of Radioactivity*, trans. Robert W. Lawson (Oxford: Oxford University Press, 1926), pp. 174–175; idem, *Lehrbuch der Radioaktivität* (Leipzig: Barth, 1931), pp. 215–216; M. Curie, *Radioactivité* (Paris: Hermann, 1935), p. 426. I. Curie, *J. Phys.*, **6** (1935), 417, in F. and I. Joliot-Curie, *Oeuvres scientifiques complètes* (Paris: Presses Universitaires, 1961), pp. 614–618. Franco Rasetti, *Elements of Nuclear Physics* (New York: Prentice-Hall, 1936), p. 201. Von Weizsäcker, *Atomkerne*, p. 6. See Flammersfeld, "Atomkern-isomerie," p. 75.

[19] L. Szilard and T. A. Chalmers, *Nature*, **135** (1935), 98; Flammersfeld, "Atomkern-isomerie," p. 76; L. Meitner, "Künstliche Umwandlungsprozesse beim Uran," pp. 24–42 in E. Bretscher (ed.), *Kernphysik: Vorträge gehalten am Physikalischen Institut der E. T. H. Zurich in Sommer 1936* . . . (Berlin: Springer, 1936), p. 41.

[20] C. F. von Weizsäcker, *Naturwissenschaften*, **24** (1936), 813–814; W. Bothe and W. Gentner, *ibid.*, **25** (1937), 284. For a review see H. J. J. Braddick in Chemical Society (London), *Ann. Rep. Prog. Chem.*, **34** (1937), 7–31 (pp. 17–21).

[21] L. Meitner, O. Hahn, and F. Strassmann, *Z. Phys.*, **106** (1937), 249–270; O. Hahn, L. Meitner, and F. Strassmann, *Chem. Ber.*, **70B** (1937), 1374–1392; Meitner, "Wege und Irrwege," p. 168.

[22] Hahn, Meitner, and Strassmann, *Chem. Ber.*, **70B** (1937), 1374–1392. For example, the "eka-rhenium" reported with 16-minute half-life must have been chiefly [101] Tc, whose chemical properties were unknown in 1938. See H. Mencke and G. Herrmann, "Was waren die 'Transurane' der dreissiger Jahre in Wirklichkeit?" *Radiochem. Acta*, **16** (1971), 119–123.

[23] Meitner, *Scientia*, **63** (1938), 13; cf. Quill, *Chem. Rev.*, **23** (1938), 126 and I. Curie and P. Savitch, *J. Phys.*, **9** (1938), 355. I. Curie and P. Savitch, *ibid.*, **8** (1937), 385–387, in Joliot-Curie, *Oeuvres*, pp. 619–620. P. Abelson, *Phys. Rev.*, **53** (1938), 211 (abstract of talk given Dec. 1937); M. L. Pool, J. M. Cork, and R. L. Thornton, *ibid.*, **52** (1937), 239–240.

[24] E. Fermi, in *Nobel Lectures: Physics. 1922-1941* (Amsterdam: Elsevier, 1965), pp. 413, 417. See also, e.g., Quill, *Chem. Rev.*, **23** (1938), 153; Rasetti, *Elements*, p. 271; J. Barton Hoag, *Electron and Nuclear Physics* (New York: Van Nostrand, 1938), p. 336; F. A. Paneth, *J. Chem. Soc.* (1937), 642–654; Diebner and Grassmann, *Phys. Z.*, **38** (1937), 419–420; R. Grégoire, *J. Phys.*, **9** (1938), 427; Ernest Rutherford, address to Science Congress in India, 1937, quoted in Mark Oliphant, *Rutherford: Recollections of the Cambridge Days* (Amsterdam: Elsevier, 1972), pp. 141–142.

[25] Meitner, Hahn, and Strassmann, *Z. Phys.*, **106** (1937), 269.

[26] G. Gamow in *International Conference on Physics, London, 1934* (Cambridge: Cambridge University Press, 1935), Vol. I, pp. 60–66 (p. 62). See Meitner, "Künstliche

Umwandlungsprozesse," p. 26; S. Flügge and G. v. Droste, *Z. Phys. Chem.*, **B42** (1939), 274–280 (pp. 274–275).

[27] Perhaps the first of these near-miss searches for alpha particles was by E. Amaldi, as he describes in "Personal Notes on Neutron Work in Rome in the 30s. . .," pp. 294–351 in Charles Weiner (ed.), *History of Twentieth Century Physics* [*Storia della fisica del XX secolo*]. Proceedings of the International School of Physics "Enrico Fermi," Course 57 (New York: Academic Press, 1977), p. 317. A. Braun, P. Preiswerk, and P. Scherer, *Nature*, **140** (1937), 682.

[28] Rutherford to Chadwick, Feb. 17, 1936, quoted in Charles Weiner with Elspeth Hart, eds., *Exploring the History of Nuclear Physics*. American Institute of Physics Conference Proceedings, 7 (New York: AIP, 1972), p. 189.

[29] For the impact of the compound nucleus see, e.g., P. B. Moon, "Slow Neutrons," pp. 198–211 in Physical Society (London), *Rep. Prog. Phys.*, **4** (1937), 200–201; Gamow, *Structure*, pp. vi–vii. For history of theory see Weiner, *Exploring*, and Roger Stuewer (ed.), *Nuclear Physics in Retrospect: Proceedings of a Symposium on the 1930s* (Minneapolis: University of Minnesota Press, 1977). I am grateful to John L. Michel for letting me have a copy of "An Historical Diagnosis of Scientific Discovery: The History of the Discovery of Nuclear Fission," unpublished seminar paper, 1975.

[30] H. Bethe and R. F. Bacher, *Rev. Mod. Phys.*, **8** (1936), 82–229, p. 171; Gamow, *Structure*, p. 202; Meitner, Hahn, and Strassmann, *Z. Phys.*, **109** (1938), 538–552 (p. 549).

[31] W. Wefelmeier, *Naturwissenschaften*, **25** (1937), 525; *Z. Phys.*, **107** (1937), 332–346; *Naturwissenschaften*, **27** (1939), 110–111. C. F. von Weizsäcker, *ibid.*, 133; von Weizsäcker, "Wilfrid Wefelmeier," *Z. Naturforsch.*, **3a** (1948), 370; Hanne Höcker, "Die Wefelmeierschen Modelle der höheren Atomkerne," *ibid.*, 365–369; Siegfried Flügge, *ibid.*, **4a** (1949), 82.

[32] N. Bohr and F. Kalckar, *Kongelige danske Videnskabernes selskab. Mathematisk-fysiske meddelelser*, **14**, No. 10 (1937). John Wheeler, "Some Men and Moments in the History of Nuclear Physics. . .," pp. 217–306 in Stuewer, *Nuclear Physics*, p. 266; Katharine Way, *Phys. Rev.*, **53** (1938), 685 (abstract).

[33] Meitner, Hahn, and Strassmann, *Z. Phys.*, **106** (1937), 270.

[34] I. Curie, H. v. Halban, and P. Preiswerk, *Compt. Rend.*, **200** (1935), 1841, in Joliot-Curie, *Oeuvres*, pp. 601–603. Cf. Gamow, *Structure*, p. 202.

[35] I. Curie, H. v. Halban, and P. Preiswerk, *J. Phys.*, **6** (1935), 367, in Joliot-Curie, *Oeuvres*, pp. 607–613, see pp. 612–613. Hahn, *Chem. Ber.*, **69A** (1936), 217–227, see pp. 221–222. Contemporaries typically gave joint credit for the "discovery," e.g., Diebner and Grassmann, *Phys. Z.*, **37** (1936), 378.

[36] Oct. 30, 1936, Rutherford Papers. Hahn was also referring to Rutherford's own ascription of the discovery to Curie.

[37] I. Curie and Paul Savitch, *J. Phys.*, **8** (1937), 385, in Joliot-Curie, *Oeuvres*, pp. 619–623, see p. 623.

[38] Otto Frisch, "Lise Meitner," *Dictionary of Scientific Biography*, Vol. IX (New York: Scribner's, 1974), p. 260. Meitner, Strassmann, and Hahn, *Z. Phys.*, **109** (1938), 538–552 (p. 550); G. v. Droste, *ibid.*, **110** (1938), 84–94. "Kuriosum": W. Gerlach, introduction to D. Hahn, *Hahn* (1979).

[39] Jan. 20, 1938, special file, Joliot-Curie Papers, Radium Institute, Paris. I am indebted to M. Bordry for assistance with these papers. See Meitner, "Wege und Irrwege," p. 168.

40 I. Curie and P. Savitch, *Compt. Rend.*, 206 (1938), 906–908, in Joliot-Curie, *Oeuvres*, p. 625.

41 F. Joliot-Curie, *Textes choisis* (Paris: Editions Sociales, 1959), p. 35.

42 *Compt. Rend.*, 206 (1938), 1643–1644, in Joliot-Curie, *Oeuvres*, pp. 626–627.

43 Meitner, "Wege und Irrwege," p. 168. Yttrium-92 has a 3.5-hour half-life and 3.5-MeV beta emission; Curie and Savitch found emission of over 3 MeV. For lanthanum-141 the values are 3.7 hours and 2.8 MeV. Both isotopes are produced abundantly in fission. Promethium may have added to the confusion. For the social context of French work see Spencer Weart, *Scientists in Power* (Cambridge: Harvard University Press, 1979), Chapters 1–3.

44 *J. Phys.* 9 (1938), 355–359, in Joliot-Curie, *Oeuvres*, pp. 628–636; *Compt. Rend.*, 208 (1939), 343–346; in *Oeuvres*, pp. 639–642, see p. 640.

45 L. Szilard, *Leo Szilard: His Version of the Facts. Selected Recollections and Correspondence*, ed. Spencer Weart and Gertrud Weiss Szilard (Cambridge: MIT Press, 1978), p. 44. M. L. Pool, J. M. Cork, and R. L. Thornton, *Phys. Rev.*, 52 (1937), 239–240. Elisabeth Rona and Elisabeth Neuninger, *Naturwissenschaften*, 24 (1936), 491.

46 P. Abelson, "A Graduate Student with Ernest O. Lawrence," pp. 22–34 in *All in Our Time: The Reminiscences of Twelve Nuclear Pioneers*, ed. Jane Wilson (Chicago: Bulletin of the Atomic Scientists, 1975), pp. 25–27; P. Abelson, *Phys. Rev.*, 55 (1939), 418. Feather, oral history interview by Weiner, 1971, AIP, pp. 120–121. See also Robley Evans, oral history interview by Weiner, 1978, AIP, pp. 52–54.

47 Meitner, Strassmann, and Hahn, *Z. Phys.*, 109 (1938), 538–552; S. Flügge, *Z. Naturforsch.*, 4a (1949), 83.

48 L. G. Cook, quoted by S. Glasstone, *Sourcebook on Atomic Energy* (New York: Van Nostrand, 1950), p. 345. O. Hahn and F. Strassmann, *Naturwissenschaften*, 26 (1938), 755–756. See R. Spence, "Otto Hahn," *Biog. Mem. FRS* (1970), 294–295; F. Strassmann, *Kernspaltung: Berlin, Dezember 1938* (Mainz: Hanns Krach, 1978), pp. 16–17. The radium isomers produced from uranium were presumed to be different isotopes than those produced from thorium.

49 Hahn to Feather, June 2, 1939, 5:21, Meitner papers, Churchill College, Cambridge. Flügge, *Z. Naturforsch.*, 4a (1949), 83.

50 Hahn, "Die 'falschen' Transurane," p. 45; Hahn, *My Life*, p. 150. See Irving, *The Virus House*; Fritz Krafft, "Lise Meitner und ihre Zeit...," *Angew. Chem.*, 90 (1978), 876–892 (p. 834); letters such as Meitner to Hahn, Dec. 5, 1938, Hahn Papers, Archives of the Max-Planck Gesellschaft, Berlin (my thanks to Rolf Neuhaus for much help in the long-distance use of these papers); Hahn to Feather, June 2, 1939, Meitner papers. Parts of the Hahn-Meitner correspondence of the period are reproduced in D. Hahn (ed.), *Otto Hahn: Erlebnisse und Erkenntnisse* (Düsseldorf: Econ, 1975), pp 75 ff., but there are unmarked omissions even within letters. See also "Die Kernspaltung," *Bild Naturwi.*, (Dec. 1978), 68–91.

51 O. Hahn, "The Discovery of Fission," *Scient. Amer.*, 198 (Feb. 1958), 76–84 (p. 80).

52 D. Hahn, *Hahn: Erlebnisse*, p. 58.

53 Hahn-Strassmann notebook, Deutsches Museum, Munich, pp. 74–77. I am grateful to Ernst Berninger for loaning me a copy.

54 Hahn-Strassmann notebook, p. 73v, undated. The most likely date is Dec. 18. The experiment is described in Hahn and Strassmann, *Naturwissenschaften*, 27 (1939), 89–

95. In that paper the notation Ba III is temporarily adopted on purpose; the first indication of a conscious shift to this is on p. 82 of the notebook, dated Jan. 5, 1939, where Ra IV is put in quotes and then appears as Ba IV. Note also that, as Krafft points out, on Dec. 15 a laboratory assistant wrote "examination of La" instead of "Ac." My point is that the ability to "see" medium-weight elements reached consciousness over some time, not in one instant discovery.

55 Anyone unsure about this should at once read Sigmund Freud, *The Psychopathology of Everyday Life.*

56 These experiments began around noon on Dec. 18; Hahn-Strassmann notebook, pp. 78v, 80−81. The letter is reproduced in D. Hahn, *Hahn: Erlebnisse,* p. 152. I am grateful to Churchill College for copies of the Hahn to Meitner letters; this one, however, was missing from the Meitner Papers. Hahn quickly recognized that if the results held, "die Trans-Urane. . . sterben. Ich weiss nicht, ob mir dies nicht doch sehr leid täte." Hahn to Meitner, Dec. 28, 1938. Meitner too was upset by the likelihood that "die ganze Arbeit der letzten drei Jahre unrichtig war." Meitner to Hahn, Jan. 1, 1939.

57 Hahn to Feather, June 2, 1939, Meitner Papers. "Gezwungen": O. Hahn, *Künstliche neue Elemente: Vom Unwägbaren zum Wägbaren. Mit einer Einführung in die Geschichte der Kernspaltung* (Berlin: Verlag Chemie, 1948), p. 10; Hahn, "Die 'falschen' Transurane," p. 46.

58 On Strassmann a work I expect to be very useful is Fritz Krafft, *Im Schatten der Sensation: Leben und Werk von Fritz Strassmann* (Weinheim: Verlag Chemie, 1980).

59 Hahn and Strassmann, *Naturwissenschaften,* 27 (1939), 11−15. On reluctance to publish see Hahn to Meitner, Mar. 3, 1939, in D. Hahn, *Hahn: Erlebnisse,* p. 155; Hahn, "The Discovery of Fission," p. 82.

60 Hahn to Meitner, Dec. 28, 1938, Meitner Papers.

61 Meitner, "Wege und Irrwege," p. 168; Frisch, "Meitner," *DSB,* Vol. IX, p. 262; Frisch, oral history interview by Weiner, 1967, AIP, p. 35. Meitner and Frisch, *Nature,* 143 (1939), 239−240. Meitner to Hahn, Jan. 1, 1939, Hahn Papers.

62 "Gesetzen der Kernphysik" to "Erfahrungen der Kernphysik," galleys with MS additions, Hahn Papers.

63 Lew Kowarski, oral history interview by Weiner, Oct. 1969, AIP, p. 74. Frisch, oral history interview by Weiner, 1967, AIP, p. 35. Abelson, "A Graduate Student," p. 29.

64 Thomas S. Kuhn, *The Structure of Scientific Revolutions* (2nd ed., Chicago: University of Chicago Press, 1970), Chapters 3−4. Kuhn himself now advocates a more complex teminology (see below, n. 72).

65 For a good bibliography see Stephen Brush, "Scientific Revolutionaries of 1905. . .," pp. 140−171 in M. Bunge and W. R. Shea, eds., *Rutherford and Physics at the Turn of the Century* (New York: Science History, 1979), Notes 4−8, pp. 163−164. I found especially useful Stephen Toulmin, *Human Understanding* (Princeton: Princeton University Press, 1972), Vol. I, pp. 106−107, 113−121; W. Berkson, "Some Practical Issues in the Recent Controversy on the Nature of Scientific Revolutions," pp. 197−209 in Robert S. Cohen and Marx W. Wartofsky, eds., *Methodological and Historical Essays in the Natural and Social Sciences.* Boston Studies in the Philosophy of Science, Vol. 14 (Boston: Reidel, 1974).

66 Hahn, *Applied Radiochemistry,* p. 33.

67 Hahn, *Künstliche neue Elemente,* p. 7.

68 Also, e.g., Hoag, *Electron and Nuclear Physics,* p. 315. Hahn, Meitner, and Strass-

mann themselves used the silicon transmutation example to instruct chemists: *Chem. Ber.* **69B** (1936), 905.

69 Livingston and Bethe, *Rev. Mod. Phys.*, **9** (1937), 291.

70 See also Meitner, "Künstliche Umwandlungsprozesse," p. 25; Diebner and Grassmann, *Phys. Z.*, **37** (1936), 369.

71 In other subfields of nuclear physics, other matters were emphasized more. For example, physicists seeking to understand nuclear forces attended more to modifications in the motion of particles that interacted with a nucleus, and were less interested in nuclear changes.

72 Kuhn, *Structure*, postscript, pp. 182–187, replaces the term "paradigm" with the term "disciplinary matrix." The components that I list would be named, in Kuhn's new terminology, "models," "symbolic generalizations," and "examplars," respectively. The paradigm I delineate is that used in uranium work; for nuclear physics as a whole, a much wider range of models, exemplars, etc. is of interest.

73 Kuhn, *Structure*, pp. 52–53, 74–76.

74 See, e.g., David Edge and Michael Mulkay, *Astronomy Transformed: The Emergence of Radio Astronomy in Britain* (New York: Wiley, 1976), pp. 387–395.

75 See, e.g., E. Rutherford, *Phil. Trans. Roy. Soc.*, A204 (1904), 169–219, in E. Rutherford, *Collected Papers* (London: Allen & Unwin, 1962), Vol. II, p. 708.

76 Mircea Eliade, *The Forge and the Crucible: The Origins and Structure of Alchemy*, trans. Stephen Corrin (New York: Harper & Row, 1962).

77 See Lawrence Badash, *Radioactivity in America: Growth and Decay of a Science* (Baltimore: Johns Hopkins University Press, 1979), pp. 25–26, 125–136; Spencer Weart, "Nuclear Fear," in preparation for Dial Press, New York.

78 For fuller discussion see Weart, *Scientists in Power*, pp. 42, 53, 81–82, 145–146 and passim.

79 Schroeder-Gudehus, "Argument for Self-Government." L. Meitner, "Looking Back," *Bull. Atom. Scient.*, **20** (Nov. 1964), 2–7 (p.6); *Z. Phys.*, **106** (1937), 250n, 257.

80 O. Hahn, "Tätigkeitsbericht der Kaiser Wilhelm-Gesellschaft," *Naturwissenschaften*, **26** (1938), 324. See also *ibid.*, **23** (1935), 414, **25** (1937), 371–372, etc. Zimen, "Einige Erinnerungen," pp. 19–80. Meitner and Delbrück, *Aufbau der Atomkerne*, p. 38. See also above, notes 9, 11.

81 Nor were the military applications wholly coincidental; the value of science in warfare was recognized when Hahn's institute and others were funded. See Weart, *Scientists in Power*, passim.

FRITZ KRAFFT

5. INTERNAL AND EXTERNAL CONDITIONS FOR THE DISCOVERY OF NUCLEAR FISSION BY THE BERLIN TEAM

My subject concerns the constellation of conditions which led to the discovery of nuclear fission by Otto Hahn, Lise Meitner, and Fritz Strassmann.[1] It is on purpose that I mention all three names of the Berlin team and that I order them alphabetically, in spite of the fact that today Hahn is usually credited with the discovery of the fission of heavy nuclei. Stressing Hahn's name, however, ignores the fact that Hahn himself called Fritz Strassmann a codiscoverer. It is absolutely wrong to attribute Hahn's attitude to modesty and to regard that as the motive that caused him to let his young assistant share his fame. I am much more convinced that it was the *cooperation* of all three scientists, including Lise Meitner, which gave rise to the discovery — at least in the way that it was done, with extremely weak radioactive samples, and as early as it was done.

In our context it is of no interest that experimental physics at that time had reached such a level — through improved spectrographic methods, for example — that it could make immediate and successful use of that discovery; nor is it relevant that physicists would have made the discovery themselves soon. It was only afterwards that those physicists who studied nuclear reactions realized with anger how close they had been to the discovery of fission. Before then, they were preoccupied with other theories and were thus not able to undertake the systematic study of the decay of atomic nuclei from a physical point of view.

But even in chemistry and nuclear (or radio-)chemistry a special constellation of internal and external conditions was necessary; and this constellation occurred only with the working team of Otto Hahn, Lise Meitner, and Fritz Strassmann, and it took place only at the Kaiser Wilhelm Institute of Chemistry in Berlin. The inquiry into this special constellation, therefore, should give priority to the question why nuclear fission was not discovered as early as 1935–1936, when the liquid-drop model of the atomic nucleus had been developed by theoretical physicists; for once the discovery was made by chemical methods, it was this model which made possible the right explanation of the cleavage called fission. At the same time answers should be given to the question why the discovery was not made by the Rome team under Enrico Fermi or by the Paris group of Irène and Frédéric Joliot-Curie.[2]

William R. Shea (ed.), Otto Hahn and the Rise of Nuclear Physics, 135–165.

The important scientific discoveries and theories relating to atomic transmutation up to the mid-thirties were known to all three groups, at least to individual members of the groups.[3] We will see that some of the theoretical assumptions actually had a hindering effect, whereas the possible consequences of others were unforeseen, both within and outside the three groups. The experimental discoveries which followed were published very quickly (most of them too quickly), and therefore the temporal advantage which one group had over the others was only very small. But nevertheless one wonders why the discovery of fission was made neither by the Fermi group, which first bombarded uranium with neutrons and first produced the alleged "transuranic elements" (which, indeed, consisted mostly of fission products), nor by the Paris group, which had discovered artificial ("induced") radioactivity and had been engaged in irradiating uranium (and thorium) with neutrons since 1934. The Berlin team in fact started with the results of the investigations of the other two: in one case clearing up the isomeric series of the alleged transuranic elements and the decay products of irradiated thorium, in the other investigating the so-called 3.5-hour substance of Irène Curie and Pavle Savitch which then led to the discovery of fission.

It would be far too simple to attribute the feat only to the extraordinary creative capacity of the Berlin team or of one of its members. On the contrary, I do believe that the work of that group was crowned with success first and foremost because it was a really interdisciplinary team. The other two groups (and the other scientists engaged in atomic transformation and decay work) consisted almost exclusively of physicists led by a physicist — in Rome by Enrico Fermi, in Paris later on by Irène Curie. In Berlin, too, it was the physicist Lise Meitner who initiated the investigations. She was the head of the physical department of the Kaiser Wilhelm Institute of Chemistry. But there was a chemical department, too, directed by Otto Hahn, the head of the entire institute. Meitner had learned from the papers of the Rome group how important chemistry was: that the scanty chemical knowledge of a physicist was just as insufficient as the occasional help of a chemist. It was she who suggested the new collaboration with Hahn after a lapse of twelve years, and it took her several weeks to persuade him to take part in the investigation of the transuranic elements.[4] But then the radiophysicist and the radiochemist became *equivalent* collaborators, just as they had been in the years 1907 to 1922. And within a short time, Fritz Strassmann, whom they included from the beginning because of his analytical training and aptitude, also became an *equivalent* collaborator.

Of course, the discovery was based neither on the sudden ingenious idea of a single person nor on a stroke of luck (as is claimed sometimes and as some of Hahn's earlier remarks seem to suggest[5]). It was based on the hard intellectual and experimental work of a well-functioning interdisciplinary team, which was certainly by no means on the beam every time, since they did not know what the end result would be. We will see, however, that their seeming detours represent an important part of the scientific or internal presuppositions of the discovery. But first we will look at some of the external presuppositions and conditions, most important of which was the composition of the team.

THE BERLIN TEAM AND ITS DELIMITATIONS

In this team Otto Hahn was the organic and nuclear chemist,[6] Fritz Strassmann was the analytical and physical chemist,[7] and Lise Meitner was the theoretical and experimental nuclear physicist.[8] The methodology of no one field was dispensable.

In agreement with Fritz Strassmann I am convinced that Lise Meitner's part even up to the finding of the barium was so great that she should be ranked among the discoverers,[9] not only because of her correct theoretical interpretation (the first), in collaboration with her nephew Otto Robert Frisch, of the results of the experiments of the Berlin chemists. It was Meitner who had instigated in the autumn of 1934 those joint studies on the "transuranics" that led into the discovery, and until her forced departure in July 1938 she had experienced all the ups and downs of the investigation, a knowledge of which was prerequisite for the later discovery. Her manner of thinking was so much integrated into the team that whatever she remarked critically in her letters was taken into account immediately.[10]

Therefore it was not only tactful regard for their old colleague, and it was not only the different political climate within the Berlin institute, which caused Hahn and Strassmann to refrain from discussing problems of theoretical physics with the physicists of Meitner's former staff or those of the neighboring Kaiser Wilhelm Institute of Physics, but to continue the discussion with *her* by the more complicated route of personal correspondence between Berlin and Stockholm. Indeed, Hahn also communicated with Meitner's nephew Frisch, who was working with Niels Bohr in Copenhagen (he had emigrated as early as 1933) and who was in close contact with his aunt.[11] On the occasion of his lecture trip to Copenhagen on November 13, 1938, Hahn also discussed with Bohr the theoretical interpretation of the products of transformation of heavy nuclei by neutron bombardment.

Bohr was then the most prominent authority in the field of atomic theory, but he was also a friend of Lise Meitner's; and she, it turned out, was in Copenhagen at the same time to spend some days working at Bohr's institute.[12]

After Strassmann and Hahn had found out that the supposed isotopes of radium had the same chemical characteristics as barium (without being able to give any theoretical explanation for this result), Hahn wrote to Meitner on December 19: "Well, consider if you can think of any possibility. If you can propose something that you can publish, then it would still be a kind of cooperation of us three."[13] At that time, indeed, Hahn still thought of Meitner as belonging to the team of three.

The physicists in Berlin and even the other chemists at the Kaiser Wilhelm Institute learned the results of the experiments only upon their publication in the first issue of *Die Naturwissenschaften* for 1939 – not earlier than January 6.[14] That is why they took it amiss that Hahn and Strassmann had not shared the details of their research and had not informed them about the discovery itself. Indeed, the Berlin physicists thought that if only they had been informed in time, they could have found the fission interpretation of the experimental results as early as or even earlier than Meitner and Frisch[15] – and it is by all means possible. But one has to consider also that the discovery was done in the last days before Christmas vacation, the paper was written on the first day of this vacation, and it was published two days after this holiday period was over.

This is the reason why none of the Berlin physicists and chemists are now able to give any information about the considerations and reflections that led to the discovery. Therefore I had to look for other sources in addition to the printed papers and books by Otto Hahn, which include but little about our question. I looked through the letters which Hahn and Meitner wrote to each other after her departure and before the discovery. This correspondence has been published (in 1975), but in parts only.[16] Nearly all the letters are preserved in the archives of Churchill College at Cambridge University and in the Max Planck Society archives in Berlin.[17] My second source was the notebooks "Chemie II" and "Chemie III" which contain the records of the experiments and measurements of Hahn and Strassmann. The first of these notebooks, preserved at the Deutsches Museum in Munich,[18] had never before been consulted apart from the two famous pages. The second notebook was kindly made available to me for inspection by Strassmann himself.[19] And he, the only member of the team that was still alive when I made my investigations, was my third source (he died in April 1980). He had several talks with me, although he was suffering greatly after two strokes in 1974. He also let me inspect some handwritten records from earlier years.[20]

It is my intention to show that the discovery was the work of the team with special contributions by all three members and that it was successful only through cooperation. (It is for this reason, therefore, that I have to try to analyze the contributions of the individual members.) It seems to me that Otto Hahn himself recognized the close collaboration as well as the separate contributions of the three scientists, realizing that even the remaining two would not have arrived at their results without their third companion. At the end of their famous paper on the detection of barium delivered on December 22, 1938, he unconsciously differentiated between their contributions when he wrote as follows:

Because of the experiments briefly outlined, as chemists we ought to change the designations ... and insert the symbols Ba, La, Ce instead of Ra, Ac, Th [that was Strassmann's point of view at that time[21]] . As "nuclear chemists" closely connected with physics in a certain way we [i.e., Hahn] cannot yet make up our mind to undertake that leap which contradicts all previous experience of nuclear physics [Meitner's influence]. A number of strange coincidences might have counterfeited our results [Hahn's habitual cautiousness].[22]

And Werner Heisenberg was certainly right when he stated in his commemorative address in 1968, relating the contributions of Otto Hahn and Lise Meitner to their earlier joint work:

It seems to me that Hahn owed his success primarily to his qualities of character. His untiring energy, his immense industry in acquiring new knowledge, and his incorruptible honesty allowed him to work even more accurately and more conscientiously, to reflect on his own experiments with even more self-criticism, and to carry out even more controls than most of the other scientists who entered the new territory of radioactivity. Lise Meitner's relation to science was somewhat different. She not only asked "What", but also "Why". She wanted to understand ... , she wanted to trace the laws of nature that were at work in that new field. Consequently her strength lay in the asking of questions and in the interpretation of experiment. We may suppose that also in their later joint work Lise Meitner exercised a strong influence on the asking of the questions and the interpretation of the experiments and that Hahn mainly felt responsible for the thoroughness and accuracy of the experiments. However, that partition did not become obvious in their joint publications.[23]

Yet here there is no mention of Fritz Strassmann, who supplied that component of analytical certainty which Hahn did not have.

Hahn had been trained in organic chemistry and was a pioneer in radiochemistry with a genius for analyzing radioactive decay curves, whereas

Strassmann had had strong training in analytical chemistry at the School of
Technology in Hannover. It is therefore not accidental that Meitner and
Hahn chose him as their collaborator when they began searching for the
transuranics, tirst detected by Fermi in 1934. And we will see that the role
of analytical chemistry ultimately was a most important one. Moreover,
Strassmann had also taken his doctorate in physical chemistry in Hannover
before he came to Berlin in 1929 to get more specialized training in chemistry
at the Kaiser Wilhelm Institute, an institute of pure research, whereas the
School of Technology was oriented toward practical applications. His profound knowledge in the field of analytical chemistry (well revealed in the
years up to 1929) and his manner of thinking as a physical chemist formed
the missing link between the radiochemist and the radiophysicist and knitted
together the methodologies of the fields of the two older, skillful scientists.
Moreover, in the meantime, Strassmann had been introduced to the techniques
of radiochemistry in special courses at the institute. These techniques for
identifying radioactive substances had been developed for the most part by
Hahn and Meitner themselves in the years from 1907 to the early 1920s
— after Hahn had learned these methods in connection with alpha radiation
at McGill University under Ernest Rutherford, the leading figure in radioactivity at that time.

When he was asked to describe the beginning of their teamwork, an
unusual activity in the 1930s, Hahn said in a television interview:

Really, we never thought of teamwork; but I already had worked together with Lise
Meitner for years [i.e., from 1907 to the early 1920s], and we had published a number
of papers. And then Strassmann came with his excellent chemistry, and then — when he
had stayed with us and had made himself acquainted [with radiochemical methods] —
we at once made up our mind to collaborate with him since we could only learn from
him. Meitner was a physicist, I was a radiochemist, and Mr. Strassmann, indeed, was just
the man we sought for good methods of chemical separation which we needed.[24]

And it was Fritz Strassmann whom Otto Hahn nominated as a unique
candidate for the Enrico Fermi Award of the U.S. Atomic Energy Commission, "for especially meritorious contributions to the development, use,
or control of atomic energy." In his letter to the chairman of the Advisory
Committee, Manson Benedict, of January 9, 1964, replying to the latter's
repeated invitation to nominate a candidate for the prize, Hahn wrote:

As I did not change my mind in the meantime, I should like to repeat what I have
written last year.
    For a selection I should like to nominate my colleague Dr. Fritz Strassmann, sharing
with me the results of the discovery of the fission of uranium. As you may perhaps

know, we have worked together for many years. Dr. Strassmann having been acquainted with the subject just as well as myself. As I have received the Nobel Prize for the year of 1944 for the discovery of uranium fission, I should think it justified if Dr. Strassmann would also receive a recognition of his very good and successful researches... [25]

The committee, however, did not follow Hahn's proposal, but chose in 1966 all three scientists of the Berlin team "to receive the award jointly because of their *combined* and individual efforts in discovering nuclear fission," [26] although one of these scientists had been the nominator himself, and although it was the first time that a non-U.S.-citizen (and a woman) was awarded the prize and the first time that the award was shared by more than one recipient.

   This constellation of fields of activity and training was but one of the, so to say, external-internal conditions of the discovery which distinguished the Berlin team from the groups in Rome and Paris. The coming together and staying together of the team was a result of many other conditions, on the one hand related to the personal careers, social origins, and characters of the three members, and on the other to the political and social situation in Germany preceding World War II.

THE PERSONAL CAREERS [27]

Otto Hahn was the son of a prosperous Frankfurt businessman who enabled him to further his education at Marburg University and his training in chemistry as a research student (*Volontär*) in London under Sir William Ramsay and thereafter under Rutherford in Montreal. More by chance, he "transmuted" — as he used to say — at this time from an organic chemist to a radiochemist and discovered two new radioelements, radiothorium and radioactinium (later explained as merely unknown isotopes of known elements). He then went to Emil Fischer, the head of the institute of chemistry at the Berlin University, in the summer of 1906. Radiochemistry was at that time a new field, largely unknown in Germany, and Fischer was very skeptical about this "new-fangled" field. He accepted Otto Hahn merely because of the good recommendations of those leading authorities Rutherford and Ramsay. Thus Fischer placed at Hahn's disposal a room in the basement of the institute that had once been used as a carpenter's workshop (*Holzwerkstatt*), where he could continue his radiochemical work.

   Because he had not been "transmuted" at Rutherford's institute to a radiophysicist, too, Hahn needed the help of a physicist for this new field, and he asked Heinrich Rubens, the head of the institute of physics, for such an assistant. However, there was nobody who had worked on radioactivity, until a young lady from Vienna came to Rubens asking for a place in his laboratories. She had already worked on radioactivity as a physicist at the University of Vienna, because the temporary successor of Ludwig Boltzmann, Stefan Meyer, had devoted his attention to this new field and had allowed her to continue her experimental work in physics with him. She had carried out measurements on the absorption of alpha and beta radiation with several materials. She came to Berlin because Boltzmann in his lectures had failed to mention the new developments in theoretical physics by Max Planck and Albert Einstein. After Boltzmann's death in 1906 she heard Planck's name for the first time. He had been invited to take Boltzmann's post and came of Vienna to look around the institute. That aroused her desire to spend several semesters at Berlin University in order to gain (as she put it) "a real understanding of physics."[28] She was allowed to attend Planck's lectures, although woman students were not welcome and were not granted admission to the universities in Prussia before 1908.[29] But besides attending theoretical lectures, Lise Meitner wanted to continue experimental work. So she applied to Heinrich Rubens and he sent her to Otto Hahn.

But there was a difficulty. Hahn worked not at Rubens's institute but instead at the institute of the chemist Emil Fischer, who, as everybody knew, cherished a strong aversion to women in his laboratories. He refused Meitner's request to work with the male students in the laboratories directed by Hahn. But there was one possibility: the *Holzwerkstatt* in the basement of the institute, where Hahn did his radiochemical research, had a detached entrance, and there she was allowed to work, provided she did not enter the other rooms of the institute. What initially appeared to be the usual kind of discrimination (although Meitner did not feel that way, since she was only interested in the possibility of experimental work) soon turned out to be a great opportunity. For this restraint resulted in the close and complementary collaboration of Meitner and Hahn,[30] which was only terminated in mid-1938 by the Nazi race laws.

Both her parents had Jewish ancestors and Meitner was therefore considered a Jew from 1933 onward. The only reason she did not fall victim of the "Law for the Re-establishment of the Professional Civil Service" of April 1933[31] was that she was an Austrian. Yet a change occurred after the annexation of Austria in early March 1938. Now Meitner became a German national, and all the German race laws were applicable to her. Close friends such as Hahn, Peter Debye, and Carl Bosch took pains to arrange her emigration, which was illegal because an order was issued by Heinrich Himmler which prohibited renowned Jews from leaving the German Reich for abroad "to act there against the interests of Germany according to their inner persuasion as representatives of the German sciences or even with their reputation and experience."[32] With the help of her Dutch colleague Dirk Coster she succeeded in escaping unnoticed in mid-July 1938. She went via Holland to Sweden, where Debye, Bosch, and Bohr had obtained a place for her to work in the Nobel Institute for Physics, recently founded at Stockholm. But the working conditions there did not bear any comparision with those at the Kaiser Wilhelm Institute of Chemistry in Berlin, where she had been head of her own department since 1917.

It had been the political attitude of Hahn, the head of the institute since 1929, which had made it possible for Meitner to remain there so long after 1933. But outside the institute she became isolated step by step. She had lost her authority to teach at Berlin University in September 1933 in consequence of the law mentioned above; she was not allowed to attend the famous Physical Colloquium of Max von Laue nor to give any address or lecture inside or outside the university. Only a few friends were left on her side. The institute itself was political divided, but apart from Kurt Hess — a National Socialist, hardly willing to make any concession, who had also betrayed the preparations of her flight — there seems to have been no hostility toward Lise Meitner. Indeed, undoubtedly it was this isolation which in the autumn of 1934 helped motivate her to persuade Hahn to embark on another joint study.

Fritz Strassmann had been in a similar position. Like Hahn, he was not willing to join the Nazi Party or any other Nazi organization, but unlike Hahn he did not have employment at the time of the seizure of power by the Hitler regime. Having gained his doctorate in physical chemistry at the Technische Hochschule of Hannover under the supervision of Hermann Braune, he had been his assistant for a short time. With a rather modest scholarship granted by the Notgemeinschaft der Deutschen Wissenschaft, he then came to the Kaiser Wilhelm Institute in the middle of 1929 in order to receive further training in radiochemical methods and thus to improve his chances for a good job in the chemical industry. It was a time of great unemployment in Germany, a situation which had prompted Strassmann to choose a school of technology (instead of a university) at which to study chemistry, not only in greater breadth but also with more emphasis on practical applications. It also prompted him to take his doctor's degree in physical chemistry, because this was what the chemical industry wanted. It was this breadth in his education and training, his knowledge, and above all, his skill at chemical analysis that caused Hahn to offer him the scholarship, and twice to file an application of continuation with the Notgemeinschaft.

When his scholarship had come to an end in September 1932, Strassmann had to support himself by preparing students for their examinations, as he had done during his undergraduate years. But he was allowed to continue his research work in Hahn's private laboratory, without pay but also without the flat rate which the research students had to pay to the institute. Later on, Meitner induced Hahn to grant Strassmann support of 50 marks per month from mid-1934, taken from a private fund he had at his disposal for special contingencies. After the seizure of power by Hitler and the National Socialists, Strassmann had to refuse the offer of lucrative posts in the chemical industry, because his entering a firm would have meant having to join the Party or a Nazi organization. This step he resisted until the collapse of the German Reich in 1945. For that reason the "politically stubborn" Strassmann was debarred from *Habilitation*, too. This qualification for a professorship most of the other members of the institute had acquired in the meantime, while Lise Meitner had been refused her right to hold academic lectures (*venia legendi*, i.e. *Habilitation*) as early as September 1933 by virtue of Paragraph 3 of the Law for the Re-establishment of the Professional Civil Service, which ordered the dismissal of civil servants of non-Aryan descent. Paragraph 4 of this law ordered dismissal for political reasons.

Even when Fritz Strassmann was employed as assistant at half pay as of January 1, 1935, after being incorporated into the working team, he should soon have left the institute — unless the political situation forced him to remain, if he wanted to work on chemical problems. But on account of his political stance he was isolated within the institute, too, and dependent on Hahn. If Strassmann had had another position at the institute, he probably would have developed his own field of work as the other members of the institute did, for teamwork was an unusual thing at that time.

Thus it was the political situation with its social consequences on the one hand, and also the liberally political attitude of the head of the institute on the other, which rendered possible the formation and adhesion of the team. In the case of the Fermi group, the political situation of Fascistic Italy separated the members of the team.

THE KAISER WILHELM INSTITUTE OF CHEMISTRY

At least one other external condition must be mentioned: the foundational support of the Kaiser Wilhelm Institute of Chemistry in Berlin-Dahlem. I cannot enter into the particulars of the long history of the preceding period;[33] however, it is essential in our context that, not the Kaiser Wilhelm Society, but the Verein Chemische Reichsanstalt and from 1921 the Emil Fischer Society for the Advancement of Chemical Research were the real foundations of the institute, itself maintained by the chemical industry. The Kaiser Wilhelm Society provided only a very small allowance. That accounts for the relative independence of the institute in relation to the Law for the Re-establishment of the Professional Civil Service, so that Meitner could stay here until mid-1938 and Strassmann could be employed even in 1935.

When the institute was established in 1912, Otto Hahn was appointed the head (and only member) of a small department for radioactivity. Lise Meitner initially joined Hahn's department as a guest and became in 1917 head of a department of her own, set up through Emil Fischer's good offices with I. G. Farbenindustrie. However, it was not only the institute's financial basis which provided them with new possibilities for research after 1912, especially when Hahn became *de jure* head of the whole institute in 1929 (*de facto* in 1926) and the other departments were dissolved; probably even more important with regard to the subsequent investigations of both natural and artificial radioactive substances was the move to a newly erected building. The *Holzwerkstatt* had become radioactively contaminated, and it would soon have been impossible to study weakly radioactive substances there. From this experience Hahn and Meitner had learned a lesson: in the new building they worked with extreme care and discipline from the very first, and so did their staffs.[34] As a result, practically no radioactive contamination was to be found in the laboratories on the ground floor (see Fig. 24), where in the first years the department of Hahn and Meitner was located, and later Meitner's department of nuclear physics, where the experiments with neutron-irradiated uranium and thorium were carried out. Here the team could work with extremely weak radioactive samples that could barely be detected by the counters available at that time, whose limit was 15 particles per minute.

These are some of the external conditions which did not exist at the institutes in Rome and in Paris where neutron irradiations of uranium and thorium had also been studied. (In Rome, for instance, irradiations, chemical operations, and measurements were carried out in one and the same room.) But what about the scientific or internal presuppositions?

THE PRODUCTION OF "TRANSURANIC ELEMENTS"

Lise Meitner and Otto Hahn were in Russia in the autumn of 1934, just at the

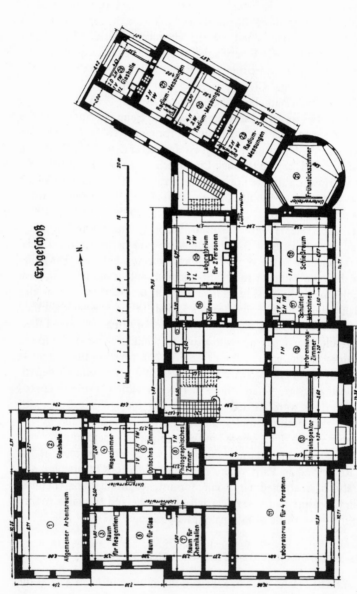

Fig. 24. The ground floor of the Kaiser Wilhelm Institute of Chemistry in Berlin-Dahlem. Rooms 23/25/27/29 were the original department of Hahn and Meitner. After 1926 the whole floor was the department of physics (Meitner's), while the department of chemistry (Hahn's) was on the first and second floor. On the ground floor the experiments with uranium and thorium were also carried out: in room 29 were the irradiations with neutrons, in room 23 the measurements; in room 20, the so-called private laboratory of Hahn, Strassmann carried out the chemical separations. (From E. Fischer and E. Beckmann, *Das Kaiser-Wilhelm-Institut für Chemie Berlin-Dahlem*, Braunschweig: Vieweg, 1913, p. 66.)

time when Enrico Fermi's famous paper on the production of transuranic elements was published.[35] Because of some inconsistencies in Fermi's statements, Meitner thought that the results should be reexamined.

Indeed, since 1932 the whole of nuclear physics had entered a new stage. In this year deuterium had been discovered by Harold C. Urey, the positron (then recognized as a nuclear component) by Carl D. Anderson, and the neutron by James Chadwick. These discoveries and the invention and improvement of the cyclotron gave new physical insights into nuclear structure. In 1934 followed the discovery of the artificial ("induced") radioactivity of atoms with relatively low atomic weight by Irène and Frédéric Joliot-Curie. That brought into Fermi's mind the idea of using, instead of the high-energy alpha particles of the Joliot-Curies, uncharged Chadwick neutrons as projectiles which should be able to penetrate even into the nuclei of highly packed atoms and to transform them in the same way that happens in natural radioactivity. If the product of this transformation is a daughter of an alpha-emitting element then, according to the displacement laws of Rutherford, or rather of Soddy and Fajans, the atomic decay should lead to an isotope whose place in the periodic table is two positions below that of the parent, that is, of the bombarded element. If it is the daughter of a beta emitter, then it takes the place of the subsequent element.

Therefore, it was obvious that Fermi's beta-emitting uranium produced by neutron irradiation led to elements beyond the last natural element, uranium; in other words, it led to transuranic elements. Fermi and his staff had been able to identify beta emitters with half-lives of 10 seconds, 40 seconds, and 13 minutes. With the help of chemical methods using manganese dioxide as the carrier they excluded lead (82), bismuth (83), and the elements from radon (86) to uranium (92) as the substances produced. Then, using rhenium sulfide as the carrying material, they tried to prove that the 13-minutes-half-life product was eka-rhenium, that is, the transuranic element with the ordinal number 93. The chemical operations were carried out by Oscar d'Agostino, who then was a research student at the Paris institute of Irène Curie but worked together with Fermi's group when he took his holidays at home in Rome. Fermi, however, did not believe that the chemical procedures used could tell him whether the transuranics had definitely been produced.

At this point Lise Meitner wanted to make use of the experience which she and Otto Hahn had had in identifying radioactive substances in earlier years, but Fermi's attempts at chemical exclusion and identification had shown that good analytical knowledge and practice were also needed. It was then that they integrated Strassmann into their team.

To get an idea of what was done, we have to consider how the periodic table was understood at the time (see Fig. 25).[36] The actinides were still

O			I						II
			H / 1						He / 2

O	I	II	III	IV	V	VI	VII	VIII
He / 2	Li / 3	Be / 4	B / 5	C / 6	N / 7	O / 8	F / 9	Ne / 10
Ne / 10	Na / 11	Mg / 12	Al / 13	Si / 14	P / 15	S / 16	Cl / 17	Ar / 18

O	Ia	IIa	IIIa	IVa	Va	VIa	VIIa	VIIIa			Ib	IIb	IIIb	IVb	Vb	VIb	VIIb	VIIIb
Ar / 18	K / 19	Ca / 20	Sc / 21	Ti / 22	V / 23	Cr / 24	Mn / 25	Fe / 26	Co / 27	Ni / 28	Cu / 29	Zn / 30	Ga / 31	Ge / 32	As / 33	Se / 34	Br / 35	Kr / 36
Kr / 36	Rb / 37	Sr / 38	Y / 39	Zr / 40	Nb / 41	Mo / 42	Ma / 43	Ru / 44	Rh / 45	Pd / 46	Ag / 47	Cd / 48	In / 49	Sn / 50	Sb / 51	Te / 52	J / 53	X / 54
X / 54	Cs / 55	Ba / 56	La Ce-Cp / 57 (58-71)	Hf / 72	Ta / 73	W / 74	Re / 75	Os / 76	Ir / 77	Pt / 78	Au / 79	Hg / 80	Tl / 81	Pb / 82	Bi / 83	Po / 84	85	Em / 86
Em / 86	87	Ra / 88	Ac / 89	Th / 90	Pa / 91	U / 92												
O	1	2	3	4	5	6	7	8	9	10	11	12	13	14	15	16	17	18

Ce / 58	Pr / 59	Nd / 60	61	Sm / 62	Eu / 63	Gd / 64	Tb / 65	Dy / 66	Ho / 67	Er / 68	Tu / 69	Yb / 70	Cp / 71	Seltene Erdmetalle

Fig. 25. Periodic table according to Andreas von Antropoff, 1934. (From J. W. van Spronsen, *The Periodic System of Chemical Elements: A History of the First Hundred Years*, Amsterdam/London/New York: Elsevier, 1969, p. 160, Fig. 59.)

unknown, since they were not discovered as such until the mid-forties, although the problem of a second series of rare earths beginning near the end of the periodic table had been discussed from the energetical standpoint since 1922. The most stable of oxidation states of actinium, thorium, protactinium, and uranium (i.e., of the elements 89 to 92) were still clearly supporting their relationship to the IIIrd, IVth, V[th], and VIth main groups. Therefore, the first transuranic element should have belonged to the VIIth main group and the following elements, as homologues of the platinoids, to the VIIIth: eka-rhenium, eka-osmium, eka-iridium, etc. After a long and intensive search, Hahn, Meitner, and Strassmann at last came to the conviction that the three beta emitters discovered by Fermi's group produced the following three isomeric series[37]:

I. $\quad _{92}U + _0n \rightarrow (_{92}U + n) \xrightarrow[\text{10 sec}]{\beta} {}_{93}\text{EkaRe} \xrightarrow[\text{2.2 min}]{\beta} {}_{94}\text{EkaOs} \xrightarrow[\text{59 min}]{\beta}$

$\quad _{95}\text{EkaIr} \xrightarrow[\text{66 hr}]{\beta} {}_{96}\text{EkaPt} \xrightarrow[\text{2.5 hr}]{\beta} {}_{97}\text{EkaAu}$

II. $\quad _{92}U + _0n \rightarrow (_{92}U + n) \xrightarrow[\text{40 sec}]{\beta} {}_{93}\text{EkaRe} \xrightarrow[\text{16 min}]{\beta} {}_{94}\text{EkaOs} \xrightarrow[\text{5.7 hr}]{\beta}$

$\quad _{95}\text{EkaIr} \xrightarrow[\text{60 days}]{\beta} ?$

III. $\quad _{92}U + _0n \rightarrow (_{92}U + n) \xrightarrow[\text{23 min}]{\beta} {}_{93}\text{EkaRe}.$

Their results were reproduced in several places around the world and the chemical identification of these products was conceded, although in particular the production and existence of so many isomers put difficulties in the way of theoretical explanation according to the liquid-drop model of the atom; for the different characteristics of the isomers were preserved even when the nuclei emitted beta and gamma rays — as the Berlin team had shown.[38] And that was incompatible with the hypothesis of different angular momentums used to explain nuclear isomerism.[39] But nobody queried the identification of the chemically corroborated decay products, and nobody referred to the possibility that the identification could have originated from erroneous assumptions, because energetic reflections had suggested since 1922 that the elements near uranium and after actinium should be a second series of rare earths, so that transuranic elements could not be homologues of rhenium and the platinoids.

The results had been chemically guaranteed with more specific operations than Fermi's group had applied, so that even Ida Noddack's general objections advanced against Fermi's chemical precipitations had become untenable.[40] Indeed, she neither reiterated her criticism nor made any experimental investigation, although she was a chemical expert on rare earths and together with her husband was the codiscoverer of the element rhenium itself. Nor did anybody else take in earnest her objections, perhaps also because she and her husband insisted on their detection of element 43 (called "Masurium" by them) which nobody was able to reproduce and which undermined considerably their scientific credibility.

Indeed, neither the presumed transuranics nor these objections led the way to nuclear fission. But the practice which the Berlin team had acquired by these collaborative investigations was a preliminary condition.[41] The results had been published in several papers, with Strassmann as coauthor from the third paper onward;[42] although he had from the beginning taken an active role in the experimental part of the research, he was now considered an equivalent collaborator. In the papers published in chemical journals or in the general journal of the Kaiser Wilhelm Society, *Die Naturwissenschaften*, it was Otto Hahn who was enumerated first, in physical journals it was Lise Meitner, whereas Fritz Strassmann as the younger (although independent) experimentalist was mentioned at the end in all cases except for one.

THE ALLEGED THORIUM PRODUCTS

After clearing up Fermi's transuranic products the Berlin team from mid-1937 began to investigate anew the products of neutron-irradiated thorium. The first results they had earlier published in mid-1935.[43] The starting point here also had been observations made by Fermi. They had found two decay series: one was induced by fast neutrons and led after alpha decay to an isotope of radium which as a beta emitter transmuted into beta-emitting actinium and thence into an isotope of thorium; the second series, induced by slow ("thermal") neutrons, began with a new beta-emitting isotope of thorium produced by neutron capture and gamma emission.[44] The results had been corroborated and supplemented by the Paris group then consisting of Irène Curie, Hans von Halban, jr., and Peter Preiswerk. Meanwhile the Rome group under Fermi and scientists in Vienna (Elisabeth Rona and Elisabeth Neuninger) had also studied the products of thorium.[45] Both series seemed to be confirmed, although the existence of alpha rays could not be proved.

The new investigations of the Berlin team now could incorporate the experience acquired with the isomeric series of the transuranics. Chemical separations and measurements of the beta radiation in the case of thorium also yielded isomeric series produced by fast neutrons. Altogether the team proved the existence of three such alleged isotopes of radium and of actinium.[46] To separate these isotopes Strassmann made use of barium and lanthanum as homologues of radium and actinium; so these isotopes seemed to be confirmed, particularly when H. Braun, P. Preiswerk, and Paul Hermann Scherrer published a brief note saying that they had detected alpha rays after irradiating thorium with neutrons.[47] However, Gottfried von Droste, assistant in Lise Meitner's department, had been asked by her to find such a proof, but had failed.[48] Only later, after the discovery of the fission of uranium and thorium, could Hahn and Strassmann show that von Droste had been right, because these alleged isotopes of radium and actinium were also products of a fission, namely of thorium, by which alpha rays do not appear.[49]

The results of the thorium investigations were published in the *Zeitschrift für Physik* in mid-1938.[50] This paper was the last one which was published jointly by all three Berlin scientists — apart from the publication of results of the investigation of a long-lived (60-day) transuranic substance started almost one year before.[51] Because the paper was published in a physical journal, naturally Lise Meitner's name takes first place; but it is remarkable that here for the first time Strassmann's name is in second place and Hahn's name is last. This arrangement ought to point out the facts of the case for each expert, for (at any rate) Hahn had been the head of the institute at which the investigation had been done. Fritz Strassmann himself corroborated my inference: these investigations had been carried out by him experimentally by special request of Lise Meitner, and Otto Hahn had not taken any part in the experiments.

This statement is not at all unimportant. After Meitner's departure, the new investigations, which eventually led to the discovery of the fission of uranium, were carried out with the same methods and theoretical presuppositions, which of course later rendered them just as incorrect as Meitner's former reflections. And these investigations were inspired by Strassmann starting from just those same reflections. What had been happening before this?

CLEARING UP THE 3.5-HOUR SUBSTANCE

Since 1934 an eager rivalry existed between the Paris and Berlin groups. Especially after the Joliot-Curies in 1935 received the Nobel Prize for the discovery of artificial radioactivity (which after all should have been given to Lise Meitner and Otto Hahn, who together had also been proposed for the prize in that year), the papers published by the Paris scientists were immediately and critically read and their results reexamined. This was also done with a paper of mid-1937, wherein Irène Curie and Pavel Savitch reported their discovery of a substance with a 3.5-hour half-life which resulted as a decay product of uranium irradiated with neutrons.[52] This substance, which apparently had escaped the notice of the Berlin team, Curie and Savitch supposed to be an isotope of thorium. Meitner was eager to repeat the experiment at once, since Strassmann at her request had searched without success for such a thorium isotope within the decay patterns of irradiated uranium as early as 1935. But even upon repeating the examination with new

experiments, no thorium was to be found in the filtrate. According to Meitner's wish, this negative result was not published but was communicated by
letter. In a subsequent paper, Curie and Savitch corroborated this message
and its contents.[53] Against that they now proposed that the 3.5-hour body
was an actinium isotope. But after the refutation of thorium, this seemed
theoretically most improbable to Meitner, because in this case the neutrons
would have had to throw out of the uranium nucleus a proton in addition
to an alpha particle. Thereupon she lost all interest in further investigations
of this "Curiosum," as the substance was then called in Berlin.

In mid-July 1938 Lise Meitner had to emigrate.[54] In mid-October a new
paper was published by Curie and Savitch, in which the substance now
seemed to be confirmed as actinium. They claimed it to be similar to the
lanthanum of the potassium lanthanum sulfate, which they used as a carrier
substance and from which it should be separable only by fractional crystallization.[55] What Otto Hahn thought about this paper as he read it quickly is
not known for certain.[56] It seems that he did not take it in earnest when
he smilingly handed it over to Strassmann in his office.[57] Nevertheless, for
the first time the Paris scientists communicated exact curves of the decay,
and the descriptions convinced Strassmann of the reality of the substance.
He therefore sought a theoretical explanation which would make compre-
hensible the results of both the Paris and the Berlin groups. Following the
example of the decay scheme of thorium just published, Strassmann put
forth the following reflections:

(1) The irradiation of uranium with neutrons leads to an *alpha*-radiating thorium. – That
would have explained why the earlier measurements, carried out with β-counters, had
turned out negatively . . . .

(2) The alpha-radiating thorium has to produce a radium.

(3) If these isotopes of radium were beta emitters, then this would decay into an
actinium and this potentially into a thorium.

(4) The Paris scientists had made use of potassium lanthanum sulfate. In this case
the radium should be at least partly within the precipitation because of the sulfate ions.

A fractionation necessarily should have produced the effects described [by Curie
and Savitch].[58]

He continued his reflections by saying: "The way chosen by us eluded sulfate
ions. We chose barium chloride as a carrier. Therefore only radium could
precipitate as a foreign substance." (Hahn confirmed again and again that
the barium chloride was suggested by Strassmann.[59])

Strassmann was able to convince Hahn of these theoretical reflections,
and the two went down at once to make a control experiment instead of

Strassmann's proceeding with the investigation of the transuranics. On October 25 Hahn gave the first account of the new experiments in a letter to Meitner:

Now, at the end of last week another paper of Curie and Savitch came out about the 3.5-hour body . . . . There are quoted curves, too. We now are going to duplicate it and believe now in its existence. According to the instructions of Curie we have found the substance, perhaps better than Curie and Savitch themselves . . . . The properties in fact seem to be strange . . . . (Perhaps even a radium isotope has to do with it. But this I say only very cautiously and privately!)[60] .

Lise Meitner answered at once on October 28 with many questions.[61] Subsequent letters partly crossed in the mail, although they arrived on the next day from Berlin to Stockholm and vice versa. At that time also the first entries were made in the preserved notebook "Chemie II."[62] It starts on November 1 with Strassmann's entry describing "a further examination of the Cu-Sa [i.e., the 3.5-hour substance of *Cu*rie and *Sa*vitch] into Ra-Ac."

On November 2 and 3 Strassmann carried out experiments which Meitner had chiefly asked for, to find out whether the substance was also produced after bombarding the uranium with slow, so-called thermal neutrons. (For the deceleration they used a cylinder of paraffin which, like most of their equipment, had been made at Meitner's suggestion in the institute's workshop.) Slow-neutron bombardment, after all, should yield transuranics as before. But the 3.5-hour substance was produced here too — which was very surprising. Thereupon in a letter of November 2 Hahn entreated Meitner to consider theoretically: "How can alpha decay probably occur also with slow neutrons, and thereby once more diverse isomers?" But for that purpose she would need more details than the few hitherto indicated, as she wrote in her reply.[63] Hahn then gave her full details in a long letter of November 5/6, but he did not wait for her answer. He already was writing the paper in which the results were published and which was delivered to *Die Naturwissenschaften* on November 8, before he went to Vienna to give a lecture.

In this paper Hahn and Strassmann supported the theory that the irradiated uranium as an alpha emitter decays into a very short-lived thorium isotope, which could not be proved because of its short half-life but which is also an alpha emitter, so that it produces an isotope of radium, which as a beta emitter decays into an actinium isotope, also a beta emitter, which leads back to thorium.[64] Altogether three isomeric series seemed to be detected; later on a fourth was added and the half-lives were calculated more exactly (they left undecided whether the isomertic series could start off with the (*n*, *α*) process or with the thorium decay):

$$
{}^{238}_{92}U + n \to \left({}^{238}_{92}U + n\right)
\begin{cases}
\xrightarrow{\alpha} {}^{235}_{90}Th? & \xrightarrow{\alpha} {}^{231}_{88}Ra\ I\ (?) \xrightarrow[< 1\ min]{\beta} {}_{89}Ac\ I \xrightarrow[< 30\ min]{\beta} Th? \\
\xrightarrow{\alpha} {}^{235}_{90}Th & \xrightarrow{\alpha} {}^{231}_{88}Ra\ II \xrightarrow[14 \pm 2\ min]{\beta} {}_{89}Ac\ II \xrightarrow[\sim 2.5\ hr]{\beta} Th? \\
\xrightarrow{\alpha} {}^{235}_{90}Th? & \xrightarrow{\alpha} {}^{231}_{88}Ra\ III \xrightarrow[86 \pm 6\ min]{\beta} {}_{89}Ac\ III \xrightarrow[several\ days?]{\beta} Th? \\
\xrightarrow{\alpha} {}^{235}_{90}Th? & \xrightarrow{\alpha} {}^{231}_{88}Ra\ IV \xrightarrow[250-300\ hr]{\beta} {}_{89}Ac\ IV \xrightarrow[< 40\ hr]{\beta} Th?
\end{cases}
$$

Thus the Curie-Savitch body turned out to be a very complex product. But the theoretical presupposition of a dual and successive alpha decay by thermal neutrons was most unsatisfactory. Therefore Meitner in her letter of November 4 suggested a reexamination to see whether the substances was composed of the jointly detected transuranics.

Meitner's suggested reinvestigation was carried out by Strassmann after Hahn's return from Vienna, but without success.[65] Further experiments followed nearly daily: "Search for short-lived Ra and perhaps Ac" (Nov. 21), "Experiment to separate Ac from short-lived Ra" (Nov. 22), "Examination for Ac from Ra with a half-life of 110 minutes" (Nov. 23), and so on.[66] Another experiment tried to detect thorium by precipitation with zircon as the carrier.[67]

The first hint of the proper direction seems to be in an experiment carried out by Strassmann on November 25, "Further experiments on the co-crystallization of 'Ra' with barium salts and search for decay products from the Ac-La precipitation".[68] Here the quotation marks around the Ra signal at least doubt about the previous interpretation. In the meantime Hahn had been in Copenhagen on November 13, where he met Bohr, Meitner, and Frisch and discussed with them the energetic problems of the double alpha decay by thermal neutrons.[69] They all denied the possibility; but none of them could give a better explanation of the experimental results. The investigation had to go on.

On November 26 **Meitner** (having heard nothing since the Copenhagen meeting) wrote: "What about the work? Do you have something further? Does a demonstrable thorium exist *before* or *behind* the three series?" (i.e., within the decay scheme U $\xrightarrow{\alpha}$ Th $\xrightarrow{\alpha}$ Ra or within the decay scheme Ra $\xrightarrow{\beta}$ Ac $\xrightarrow{\beta}$ Th). For the time being, Hahn could report no new results.

On Decmber 7 and 8 Fritz Strassmann carried out further experiments with the long-lived isotopes by fractional precipitation of some barium chlorides used before plus a newly produced barium bromide, connected together through barium carbonate to barium bromide.[70] It was his aim to increase the weak radium in the preparations; since the beta rays were very soft (i.e., they had only a short range), most of the radiation was absorbed by the thick layer of the barium chloride itself.[71] (Again Hahn corroborates in his printed reports the fact that the bromide was proposed by Strassmann.[72]) He had chosen it because in the mixed crystal of the bromide the radium is more concentrated — in the ratio of 6 : 1 to barium — than in the chloride, where the ratio of radium to barium is only 1.8 : 1.

But even this enrichment of the radium failed to give a clarification. As much as ever there were great difficulties in separating the alleged active radium from the inactive barium by fractional crystallization. In addition, the degree of the radiation (i.e., of the enrichment) was not the same as they were familiar with. They considered that perhaps the few atoms of the separated artificial isotopes of radium could behave differently compared with the ponderable quantities which had been studied up to that time. Thereupon Hahn made a test in the first floor of the institute, where his own department was located and where all the work with intense natural radioactivity was done. In cooperation with Gerhard Radoch, he diluted some natural radium down to the intensity of the preparations which were separated by Strassmann from the neutron-irradiated uranium with barium chloride as the carrier. But within the fractional crystallization the few atoms of the natural radium enriched just as well and in the same ratio as ponderable quantities did. Therefore the *quantity* could not be the reason that the alleged artificial radium could not be separated from the barium carrier. On the contrary, the artificial radium ought to display the same chemical behavior as the barium.

THE DISCOVERY OF FISSION

Indeed, it seems that as a consequence of these results the names of the carrier elements, barium and lanthanum, came to be used rather unconsciously instead of radium and actinium, for Irmgard Bohne, who occasionally helped as laboratory assistant, put down in the notebook under December 15 "Examination for La from 16 minutes Ra" instead of "examination for actinium," and Hahn repeated that later, when he noted the evaluation.[73] Lanthanum indeed could not have resulted from radium, only from barium. Therefore Hahn and Strassmann started the famous indicator experiment in the evening of December 16, a Friday.[74] On Monday (December 19) Hahn wrote in a letter to Lise Meitner:

In the meantime I do work, as far as I find time to work,[75] and Strassmann does untiringly work at the uranium substances, helped By Lieber and Bohne. It will soon be eleven o'clock in the evening; at a quarter to twelve Strassmann will come again, so that I can go homeward gradually. Of course, there is something with the "radium isotopes," which is remarkable to such a degree that in the meantime we tell it only to you. The half-life periods of the three isotopes are tolerably exactly ascertained. They are separable from all the elements except barium. All the reactions are all right — except one: The fractionation does not function. Our radium isotopes behave like barium. We do not get a clear enrichment with barium bromide or chromate etc. Now last week I had fractionated ThX [i.e., Ra 227] in the first floor. That functioned as it ought to. Then on Saturday [Dec. 17] Strassmann and I fractionated one of our [alleged] "radium" isotopes with Msth I [i.e., Ra 228] as indicator. The mesothorium was enriched as prescribed, our radium not. ... We more and more come to the terrible conclusion that our radium isotopes do not behave like radium, but like barium . . .

In the case of this important and decisive experiment Hahn of course insisted on describing the experiment himself in the notebook, although only approximately and by way of later addition, when the measurements proved that the alleged radium really was barium, as is indicated by "Ba III 86 Min." instead of "Ra III" in the line before the last one.[76] The enrichment of the radium-228 in the barium chloride had the right proportion of 6 : 1; the enrichment of the alleged radium yielded a proportion of nearly 1.2 : 1. The latter therefore was not radium but radioactive barium.[77]

The paper which communicated these results was finished by Hahn on Thursday, December 22, 1938, the first day of Christmas vacation (the last pages with the new results were typewritten by Irmgard Bohne, the laboratory assistant, because the secretary was on vacation). It was delivered to *Die Naturwissenschaften* the same day.[78] The carbon copy (without the most important page 11 and without any curve) was not received by Lise Meitner until December 30,[79] because she had left Stockholm and was staying at Kungälv near Göteborg, spending the Christmas holiday with her nephew Otto Robert Frisch.

On the galley proofs of the paper, Hahn, without consulting Strassmann, who had been away for the Christmas holiday, added on December 28 a "fantasy," as he called it in his letter of this day to Meitner — that is, an attempt to interpret the results of the chemical experiments.[80] He supposed that the other fragments of the uranium atoms besides barium and lanthanum should be one of the elements 43 to 46, taking the atomic weight of the elements as a basis and supposing that these were the "transuranics" detected and proved earlier, which then should not be the higher but the lower homologues of rhenium and the platinoids. But Meitner and Frisch immediately understood that the splitting was energetically explicable only on the basis of the ordinal number of the elements, which amounts to the electrical charges of an atom. On January 1 at noon Meitner wrote in a letter to Hahn: "perhaps it can energetically be possible that such a heavy nucleus bursts"; and therefore she could not accept Hahn's "fantasy," which seemed to her to be impossible for several reasons. She would be anxious to hear Strassmann's opinion. And, indeed, Hahn had to confess that he was hasty. On January 5 he wrote to her: "Strassmann, who came back today, is emphatically more doubtful, for he pointed out to me that also the ruthenium is as likely as the osmium to volatilize at low temperature, so that our 'EkaOs' had to be volatile from solutions, what we never found." This little episode also shows that the analytical knowledge of Strassmann was valued greatly within the team.

In his letter to Hahn of January 4 Frisch for the first time intimated the explanation.[81] The results of their correct theoretical reflections, which led to the existence of the inert gases krypton and xenon as the other fragments of the splitting of uranium or thorium atoms, Hahn learned about on January 24 from the copy of their joint paper which Meitner sent to him after it was accepted by the journal *Nature*.[82] Thereupon Hahn and Strassmann immediately tried to identify these inert gases within the products of the splitting of excited uranium, now called fission at Frisch's suggestion. This was a difficult problem, because the inert gases do not combine with other chemical elements, so that none of these could be used as a carrier to extract them from the irradiated uranium salt. Strassmann therefore thought up a novel experimental arrangement in which an air stream sucks the radioactive gases into a glass tube filled with wadding, which collects the solid decay products of the beta-radiating active inert gases.[83] The evidence of the radioactive inert gases followed from the chemically proved existence of active isotopes of strontium and cesium resulting from the gases.

Thereby the fission of heavy nuclei was proved definitely.[84] The proceedings which led to it via many detours show that its discovery could have been made only by unprejudiced, critical chemists and only under certain external and internal conditions. The history of the events also shows that theoretical physics was required to give the correct interpretation of the chemical results. Unfortunately, the ideal team for that purpose — consisting of the radiochemist Otto Hahn, the analytical and physical chemist Fritz Strassmann, and the nuclear physicist Lise Meitner — had been separated shortly before the last part of the way began. But it could also have been the case that Meitner from the first had condemned the erroneous conceptions of Strassmann, which in fact led to the radium from which the barium finally

arose. Then her departure would have been an additional prerequisite for
the discovery of fission (in the way that it was done); for she could have
positively influenced the investigation with her criticism only after it had
been initiated, because she was not present to hinder the investigation of that
theoretical absurdity, which, nevertheless, within a few hours revealed findings
not to be disavowed.

CONCLUSION

The discovery of fission was an extremely complex event. It shows that
sometimes it is quite fruitful to make nonsense, but it has to be done system-
atically. Of course, if you consider the weakness of the available neutron
sources and the fact that only the very rare isotope uranium-235 is fissionable
(as was calculated by Niels Bohr and John A. Wheeler shortly after the
discovery[85]), it was a piece of luck that the false theoretical reflections
of Strassmann had led to the homologues of barium and lanthanum, so that
they were used as carrier substances. (Isotopes of barium and lanthanum
also are the most numerous fission products which have sufficiently long
half-lives.) But these fortunate circumstances existed for other scientists
as well. The Paris group had used lanthanum as a carrier, and Irène Curie
also had been conversant with radiochemical methods (which had been
developed by her mother, no less) and with the analysis of measurement
curves. But her collaborators changed frequently, whereas the social and
political situation held the Berlin team together for many years. But the most
important difference between these two groups was that only the Berlin team
included an analytical chemist (and moreover an excellent one). Of course,
the development and use of appropriate methods of chemical separation
decided the matter.

Johannes Gutenberg-Universität Mainz

NOTES

[1] For more details see F. Krafft, *Im Schatten der Sensation: Leben und Wirken von
Fritz Strassmann, dargestellt nach Dokumenten und Aufzeichnungen* (Weinheim: Verlag
Chemie, 1981); Ch. 3 includes the texts of all unprinted documents known to me which
relate to the discovery of fission and its prehistory (notebooks, letters, handwritten
records of Strassmann, etc.) and facsimiles of the corresponding papers by Hahn, Meitner,
and Strassmann. A good new sourcebook with 21 papers on nuclear chemistry and
physics from Fermi's transuranics to the chain reaction (1934–1952) is H. Wohlfarth

(ed.), *40 Jahre Kernspaltung: Eine Einführung in die Originalliteratur* (Darmstadt: Wissenschaftliche Buchgesellschaft, 1979).

[2] For the Rome group see E. Segrè, *Enrico Fermi, Physicist* (Chicago/London: University of Chicago Press, 1970); for the Paris group, M. Goldsmith, *Frédéric Joliot-Curie: A Biography* (London: Lawrence and Wishart, 1976).

[3] Cf. the synopsis in Krafft, *Im Schatten der Sensation*, Chapter 1.3.2, where a distinction is made between those discoveries, inventions, and theories (hypotheses) which were "present" and known to Hahn, Strassmann, and Meitner and those which were not.

[4] See Meitner's letter to Max von Laue, Sept. 4, 1941, quoted in K. E. Boeters and J. Lemmerich (eds.), *Gedächtnisausstellung zum 100. Geburtstag von Albert Einstein, Otto Hahn, Max von Laue, Lise Meitner in der Staatsbibliothek Preussischer Kulturbesitz, Berlin, vom 1. März – 12. April 1979* (Bad Honnef: Physik Kongress-Ausstellungs- und Verwaltungsgesellschaft, 1979), p. 116; and L. Meitner, "Wege und Irrwege zur Kernenergie," *Naturw. Rund.*, 16 (1963), 167–169.

[5] E. g., O. Hahn, "Die Auffindung der Kernspaltung," in W. Bothe and S. Flügge (eds.), *Nuclear Physics and Cosmic Rays* (= FIAT Review of German Science 1939–1946, Vol. 13; Wiesbaden: Dieterich, 1948), Pt. 1, pp. 171–178; reprinted in Wohlfarth, *40 Jahre Kernspaltung*, pp. 304–311.

[6] For Hahn see esp. his autobiographical books *Vom Radiothor zur Uranspaltung* (Braunschweig: Vieweg, 1962), translated into English by W. Ley as *Otto Hahn, A Scientific Biography* (New York: Scribner's, 1966); and *Mein Leben* (Munich: Bruckmann, 1968). Furthermore, E. H. Berninger, *Otto Hahn in Selbstzeugnissen und Bilddokumenten* (Reinbek bei Hamburg: Rowohlt, 1974); D. Hahn (ed.), *Otto Hahn, Begründer des Atomzeitalters: Eine Biographie in Bildern und Dokumenten* (Munich: List, 1979).

[7] In addition to Krafft, *Im Schatten der Sensation*, see F. Krafft, "Ein Leben im Dienste der Chemie und des akademischen Nachwuchses: Prof. Dr. -Ing. Fritz Strassmann zum fünfundsiebzigsten Geburtstag," *Jahrb. Ver. "Freunde der Universität Mainz,"* 25/26 (1976/77), 226–230.

[8] See O. R. Frisch, "Lise Meitner, 1878–1968, Elected For. Mem. R. S. 1955," *Biogr. Mem. FRS*, 6 (1970), 405–420; F. Krafft, "Lise Meitner und ihre Zeit – Zum hundertsten Geburtstag der bedeutenden Naturwissenschaftlerin," *Angew. Chem.*, 90 (1978), 876–892, English version *Angew. Chem., International Edition in English*, 17 (1978), 826–842; and Krafft, *Im Schatten der Sensation*, Chapter 2 (here also the more recent literature is quoted).

[9] F. Strassmann, *Kernspaltung – Berlin, Dezember 1938* (Mainz: privately printed, 1978), esp. pp. 18–20, 23, quoted also in Krafft, *Im Schatten der Sensation*, pp. 203–211.

[10] Strassmann, *Kernspaltung*; the correspondence between Hahn and Meitner edited (incompletely and with sentences and passages omitted without proper notation) by D. Hahn, ed., *Otto Hahn: Erlebnisse und Erkenntnisse* (Düsseldorf/Vienna: Econ, 1975), pp. 75–129, includes only the letters from Nov. 26, 1938, to Apr. 22, 1939. Passages from previous and later letters which are essential for the prehistory and early history of the discovery of fission, as well as letters omitted or edited incompletely by Dietrich Hahn, are quoted (mostly in extracts) by Krafft, *Im Schatten der Sensation*, Chapter 3.

[11] See now O. R. Frisch, *What Little I Remember* (London: Cambridge University Press, 1979).

[12] Hahn does not mention that anywhere; but see the visitor's book in the archives of the Niels Bohr Institute of Copenhagen University (a copy of the corresponding page was kindly given to me by E. Rüdiger): "L. Meitner November 10–17, 1938; O. Hahn November 13, 1938." All that Hahn says or writes anywhere is "in autumn 1938."

[13] Archives of Churchill College, Cambridge. Hahn's letters are mostly handwritten. Meitner's from mid-1940 are mostly typewritten (in which case a carbon copy also remained in her literary estate, now in the archives of Churchill College; her original letters are located in the archives of the Max Planck Society, Berlin). Copies of these letters were kindly given to me by Mrs. P. Bradford (Cambridge) and R. Neuhaus (Berlin).

[14] O. Hahn and F. Strassmann, "Über den Nachweis und das Verhalten der bei der Bestrahlung des Urans mittels Neutronen entstehenden Erdalkalimetalle," *Naturwissenschaften*, 27 (1939), 11–15 (received Dec. 22, 1938, published Jan. 6, 1939).

[15] See Krafft, *Im Schatten der Sensation*, Chapter 1.3.2 (pp. 105 ff.)

[16] See n. 10 above.

[17] See n. 13 above.

[18] Library of the Deutsches Museum, Munich, Sondersammlungen, Access. No. 1960–15. E. H. Berninger kindly permitted me to make a copy.

[19] For quotations see n. 1. The notebook "Chemie III" (beginning with the circuit experiment, Jan. 17, 1939) and the following notebooks are in the personal estate of F. Strassmann, Mainz, whereas "Chemie I" (beginning after L. Meitner's departure) seems to be lost.

[20] Now partly published in Strassmann, *Kernspaltung*, and/or Krafft, *Im Schatten der Sensation*.

[21] Strassmann as early as 1936 once had found radioactive barium instead of the alleged eka-rhenium, so, in contrast to Hahn, he was very sure in his finding of barium and lanthanum later. See Krafft, *Im Schatten der Sensation*, pp. 220 f.

[22] Hahn and Strassmann, "Über den Nachweis . . . ," p. 15.

[23] W. Heisenberg, "Gedenkworte für Otto Hahn und Lise Meitner", *Orden pour le mérite für Wissenschaften und Künste: Reden und Gedenkworte*, 9 (1968/69), 111–119 (pp. 113 f.).

[24] This interview of 1967 was repeated in the television documentary by A. Wischnewski, "Geboren in Boppard . . . Fritz Strassmann," first telecast by the IIIrd program of the Südwestfunk (Baden-Baden, West Germany) on Nov. 1, 1977.

[25] Carbon copy in the archives of the Max Planck Society, Berlin (literary estate of O. Hahn).

[26] Official announcement of the U.S. Atomic Energy Commission No. J-185, Aug. 5, 1966.

[27] See notes 6–8, and for Meitner also Frisch, *What Little I Remember*.

[28] L. Meitner, "Looking Back," *Bull. Atom. Sci.*, 20/No. 10 (Nov. 1964), 2–7 (p. 4).

[29] See Krafft, "Lise Meitner," pp. 877 f., 881 (English version, pp. 827 f., 830 f.).

[30] This close and successful collaboration also continued after Fischer lifted his restriction when the attendence of women at universities was officially sanctioned in Prussia and in the Prussian city of Berlin in 1908/09.

[31] "Gesetz zur Wiederherstellung des Berufsbeamtentums," *Reichsgesetzblatt*, 1933, Pt. I, No. 34, of Apr. 7, 1933, pp. 175 ff.

[32] See the letter of the Ministry of Education to C. Bosch, June 16, 1938, quoted in Krafft, "Lise Meitner," p. 886 (English version, p. 835).

[33] See, e.g., L. Burchardt, *Wissenschaftspolitik im Wilhelminischen Deutschland: Vorgeschichte, Gründung und Aufbau der Kaiser-Wilhelm-Gesellschaft zur Förderung der Wissenschaften* (Göttingen: Vandenhoeck & Rupprecht, 1975); E. Fischer and E. Beckmann, *Das Kaiser-Wilhelm-Institut für Chemie in Berlin-Dahlem* (Braunschweig: Vieweg, 1913).

[34] See Krafft, *Im Schatten der Sensation*, Chs. 1.3.2 and 3.2.2.1.

[35] E. Fermi, "Possible Production of Elements of Atomic Number Higher than 92," *Nature*, 133 (1934), 898 f. (reprinted in Wohlfarth, *40 Jahre Kernspaltung*, pp. 55–58); then E. Fermo, E. Amaldi, O. d'Agostino, F. Rasetti, and E. Segrè, "Artificial Radioactivity Produced by Neutron Bombardement," *Proc. Roy. Soc.*, 146 (1934), 483–500.

[36] See J. W. van Spronsen, *The Periodic System of Chemical Elements: A History of the First Hundred Years* (Amsterdam/London/New York: Elsevier, 1969), p. 160 (Fig. 59).

[37] O. Hahn and L. Meitner, "Über die künstliche Umwandlung des Urans durch Neutronen," *Naturwissenschaften*, 23 (1935), 37 f.; II. Mitteilung, *ibid.*, 230 f.; O. Hahn, L. Meitner, and F. Strassmann, "Einige weitere Bemerkungen über die künstlichen Umwandlungen beim Uran," *ibid.*, 544 f.; L. Meitner, and O. Hahn, "Neue Umwandlungsprozesse bei Bestrahlung des Urans mit Neutronen," *ibid.*, 24 (1936), 158 f. (p. 159: "Zum Schluss möchten wir noch Herrn Dr. F. Strassmann für seine ausserordentlich wertvolle Mitarbeit bei den chemischen Trennungen herzlich danken"); O. Hahn, L. Meitner, and F. Strassmann, "Neue Umwandlungs-Prozesse bei Neutronen-Bestrahlung des Urans; Elemente jenseits Uran," *Ber. Deut. Chem. Ges.*, 69 (1936), 905–919; L. Meitner, O. Hahn, and F. Strassmann, "Über die Umwandlungsreihen des Urans, die durch Neutronenbestrahlung erzeugt werden," *Z. Phys.*, 106 (1937), 249–270; O. Hahn, L. Meitner, and F. Strassmann, "Über die Trans-Urane und ihr chemisches Verhalten," *Ber. Deut. Chem. Ges.*, 70 B (1937), 1374–1392; O. Hahn, L. Meitner, and F. Strassmann: "Ein neues langlebiges Umwandlungsprodukt in den Trans-Uranreihen," *Naturwissenschaften*, 26 (1938), 475 f.

[38] Meitner, Hahn, and Strassmann, "Über die Umwandlungsreihen des Urans."

[39] C. F. von Weizsäcker, "Neuere Modellvorstellungen über den Bau der Atomkerne," *Naturwissenschaften*, 26 (1938), 209–217 and 225–230 (pp. 229 f.).

[40] I. Noddack, "Über das Element 93," *Angew. Chem.*, 47 (1934), 653 f., reprinted in Wohlfarth, *40 Jahre Kernspaltung*, pp. 60–64.

[41] See Hahn's letter to Meitner of Jan. 16, 1939 (cf. n. 13): "Es steht in unseren Uran- und Thorarbeiten [see notes 37, 43, and 50] so viel gute Arbeit, dass Strassmann und ich ohne diese Erfahrung jetzt sicher nicht schnell weiterkämen... " (D. Hahn, *O. Hahn*, omits this letter), and her answer of Jan. 18, 1939: "Was Du über die Uran- und Thorarbeiten schreibst, ist genau richtig, und ich hatte mir inzwischen auch [etwas] in diesem Sinne überlegt. Und vielleicht wäre es ganz nett, wenn Ihr das in Euerer nächsten Notiz ganz sachlich irgendwie zum Ausdruck brächtet, *nicht* etwa als Rechtfertigung für unsere früheren Arbeiten [that means the "transuranics," which to her mind ought to be withdrawn by all three together], die brauchen wir nicht; sondern nur als Hinweis, dass ohne die vorangegangenen Resultate und die Ausarbeitung der chemischen und physikalischen Arbeitstechnik die Aufklärung der Curie-Savitch[-]Beobachtungen kaum so schnell möglich gewesen wäre .... " Thereupon Hahn added at the end of the paper he and Strassmann were preparing (see n. 49): "Dass die im Vorstehenden

beschriebenen, zahlreichen neuen Umwandlungsprodukte sich in verhältnismässig kurzer Zeit mit – wie wir glauben – erheblicher Sicherhejt feststellen liessen, war nur möglich durch die Erfahrung, die wir bei den früheren, in Gemeinschaft mit L. Meitner durchgeführten systematischen Versuchen über die Transurane und die Thorumwandlungen sammeln konnten." See Krafft, *Im Schatten der Sensation*, Ch. 6.2.2.

[42] See n. 37 above.

[43] O. Hahn and L. Meitner, "Die künstliche Umwandlung des Thorium durch Neutronen: Bildung der bischer fehlenden radioaktiven 4n + 1-Reihe," *Naturwissenschaften*, **23** (1935), 320 f.

[44] See also O. Hahn and L. Meitner (with experimental collaboration of F. Strassmann), "Künstliche radioaktive Atomarten aus Uran und Thor," *Angew. Chem.*, **49** (1936), 127 f.

[45] See the specifications in the article by Meitner, Strassmann, and Hahn given in n. 50 below.

[46] See Ernst Berninger's chapter on the discovery of Uranium Z. Uranium Z was discovered by Hahn in 1922. It was the only really nuclear-isomeric isotope known until the end of the 1930s; but this discovery gave the example of the alleged isomeric series of transuranic elements and thorium products.

[47] *Nature*, **140** (1937), 682.

[48] G. von Droste, "Über Versuche eines Nachweises von α-Strahlen während der Bestrahlung von Thorium und Uran mit Radium+Beryllium-Neutronen," *Z. Phys.*, **110** (1938), 84–94 (received May 25, 1938).

[49] See O. Hahn and F. Strassmann, "Nachweis der Entstehung aktiver Bariumisotope aus Uran und Thorium durch Neutronenbestrahlung; Nachweis weiterer aktiver Bruchstücke bei der Uranspaltung," *Naturwissenschaften*, **27** (1939), 89–95 (p. 93) (communicated Jan. 28, 1939).

[50] L. Meitner, F. Strassmann, and O. Hahn, "Künstliche Umwandlungsprozesse bei Bestrahlung des Thoriums mit Neutronen; Auftreten isomerer Reihen durch Abspaltung von α-Strahlen," *Z. Phys.*, **109** (1938), 538–552 (received May 16, 1938).

[51] See Hahn, Meitner, and Strassmann, "Ein neues langlebiges Umwandlungsprodukt," communicated to *Naturwissenschaften* July 12, 1938, two days before Meitner left Germany.

[52] I. Curie and P. Savitch, "Sur les radioéléments formés dans l'uranium irradié par les neutrons," *J. Phys. Radium*, **8** (1937), 385–387.

[53] I. Curie and P. Savitch, "Sur la radioélément de période 3,5 heures formé dans l'uranium irradié par les neutrons," *Compt. Rend.*, **206** (1938), 906–908. See Meitner's letter to Hahn of Jan. 24, 1957, quoted in Krafft, *Im Schatten der Sensation*, p. 206, n. 9. Concerning the letter of Meitner and Hahn to I. Curie of Jan. 20, 1938, see the contribution of Spencer Weart in this volume.

[54] The date given by Hahn (July 17, 1938) is to be corrected to July 14, because he could confirm the receipt of the news (by telegram) of her safe arrival at Groningen by a postcard to the Coster family of July 15, 1938 (Archives of Churchill College); see Krafft, *Im Schatten der Sensation*, pp. 175 f., "Lise Meitner," p. 886.

[55] I. Curie and P. Savitch, "Sur les radioéléments formé dans l'uranium irradié par les neutrons, II," *J. Phys. Radium*, **9** (1938), 355–359 (p. 356: "Dans l'ensemble, les propriétés de R 3,5 h sont celles du lanthane, dont il semble jusqu'ici qu'on ne puisse le séparer par fractionnement").

56 That could be confirmed by Leo Yaffe, chairman of the session on the discovery of fission of this conference, who communicated that Leslie G. Cook (then working on his thesis at the Kaiser Wilhelm Institute of Chemistry under the supervision of Hahn) had reported that Hahn that morning came into the laboratory waving the issue of the journal containing the paper of Curie and Savitch and exclaimed: "Die Dame ist verrückt!" See also S. Glasstone, *Sourcebook on Atomic Energy* (New York: Van Nostrand, 1950), p. 345.

57 See also R. Jungk, *Heller als tausend Sonnen: Das Schicksal der Atomforscher* (Reinbek bei Hamburg: Rowohlt, 1964; first published Bern/Stuttgart: Scherz, 1963), pp. 68 f.

58 Strassmann, *Kernspaltung*, p. 17.

59 See, e.g., Hahn, *Vom Radiothor*, pp. 129 f.; for other references see Krafft, *Im Schatten der Sensation*, n. 23 of Chapter 1.3.1.1.

60 Archives of the Churchill College (see n. 13).

61 Archives of the Max Planck Society, Berlin (see n. 13). This letter as well as the following ones and the descriptions in the notebooks are quoted chronologically in Krafft, *Im Schatten der Sensation*, Ch. 3.2.

62 See n. 18 above.

63 Meitner to Hahn, Nov. 4, 1938 (see n. 61 above).

64 O. Hahn and F. Strassmann, "Über die Entstehung von Radiumisotopen aus Uran durch Bestrahlen mit schnellen und verlangsamten Neutronen," *Naturwissenschaften*, 26 (1938), 755 f. (received Nov. 8, 1938, published Nov. 18, 1938); for the following four isomeric series see the paper of Dec. 22, 1938 (n. 78 below).

65 Nov. 12, 1938; see notebook "Chemie II," fol. 18^V (see n. 61 below).

66 Nov. 21, 1938, "Chemie II," fols. 29^r ff.; Nov. 22, fols. 32^r ff.; Nov. 23, fols. 34^V ff. (Strassmann).

67 Nov. 15, 1938, "Chemie II," fols. 24^V ff. (Strassmann); see Meitner's letter to Hahn of Nov. 26, 1938 (below). Thereupon repeated on Dec. 2, 1938; see fol. 56^r (Strassmann).

68 F. Strassmann, "Weitere Versuche über die Mischkristallbildung des 'Ra' mit Ba-Salzen u[nd] Suche nach Folgeprodukten aus Ac-La-Niederschlag," "Chemie II,"fols. 40^V ff.; "Weiterverarbeitung", fols. 45^r ff. (facsimile of pp. 40^V f. in Krafft, *Im Schatten der Sensation*, p. 248).

69 See n. 12 above.

70 "Chemie II," fols. $64^r/63^V/64^V$ (Strassmann).

71 See, e.g., Hahn, "Die Auffindung der Kernspaltung" (n. 5 above).

72 See n. 59 above.

73 "Chemie II," fol. 74^V (I. Bohne, undoubtedly dictated by Strassmann or Hahn): "Prüfung auf La aus 16' Min [uten-] Ra" (facsimile of pp. 74^V f. in F. Krafft, "Ein frühes Beispiel interdisziplinärer Teamarbeit: Zur Entdeckung der Kernspaltung durch Hahn, Meitner und Strassmann," *Phys. Bl.*, 36 (1980), 85–89 and 113–118; p. 115). In the morning of the same day she still had written (fol. 73^V): "Prüfung auf kurzlebiges Ac." Within the day of Dec. 15, 1938, therefore, the idea of fission certainly arose.

74 "Chemie II," fols. 75^V ff. (Hahn, measurements by Strassmann and Hahn, later on also by I. Bohne and Clara Lieber). The indicator experiment was started by Strassmann, while Hahn in the forenoon stayed at the fiscal office for the sake of Lise Meitner's matter; see the entry of Dec. 17 in his pocket almanac, printed in facsimile by D. Irving,

Der Traum von der deutschen Atombombe (Reinbek bei Hamburg: Rowohlt, 1969; translation from the English: *The Virus House*, London: Kimber, 1967), p. 23.

[75] Besides the affairs of Lise Meitner on which he had to spend much time at the offices and besides his duties as the director of the institute, as member of the Prussian Academy, and as university lecturer, he at that time spent several hours every day visiting his wife Edith in the hospital. The passages in the letters referring thereto are mostly omitted "as private" by D. Hahn (see n. 10 above).

[76] See n. 74, and the facsimile in Spencer Weart's chapter (p. 111, Fig. 13). That this entry by Hahn was made as a later addition explains also why fol. 77^V, which should contain the description of the "Indikatorversuch Ac IV-Msth 2," remained blank.

[77] "Chemie II," fol. 77^r.

[78] O. Hahn and F. Strassmann, "Über den Nachweis und das Verhalten der bei der Bestrahlung des Urans mittels Neutronen entstehenden Erdalkalimetalle," *Naturwissenschaften*, 27 (1939), 11–15 (received Dec. 22, 1938; published Jan. 6, 1939, see also n. 80 below). This paper continued the paper of Nov. 8, 1938 (see n. 64 above); it is reprinted several times, e.g. in Wohlfarth, *40 Jahre Kernspaltung*, pp. 65–76, and Krafft, *Im Schatten der Sensation*, pp. 255–259 and 266f. (facsimiles), where also the corrections, modifications, and supplements within the galley proofs are quoted (see n. 80 below). The new results here only were supplemented at the end; the original title runs: "Über den Nachweis und das Verhalten der bei der Bestrahlung des Urans mittels Neutronen entstehenden Radium-Isotope"; see Hahn, *Vom Radiothor*, p. 131.

[79] Meitner's postcard to Hahn of Dec. 30, 1938 (see n. 13 above).

[80] See Hahn's letter to Meitner, Dec. 28, 1939: "Und nun kommt meine neue Phantasie. Wir haben *nicht* bewiesen, dass die Trans-Urane *nicht* Ma [i.e., element 43, technetium], Ru, Rh, Pd sind. Wenigstens weiss ich persönlich so wenig über diese Körper, dass ich sie chemisch nicht ausschliessen kann. (Strassmann ist verreist.) [This last sentence D. Hahn (see n. 10 above) omits!] Wäre es möglich, dass das Uran 239 [*sic*! see text belonging to n. 85 below] zerplatzt in ein Ba und ein Ma? Ein Ba 138 und ein Ma 101 ergäbe 239." The addition in the paper consists of the section before the last three running: "Was die 'Trans-Urane' anbelangt . . . ergibt 239!" (See n. 78 above). Meitner received the proofs on Jan. 3, 1939 (see her letter to Hahn of this day), also without the curves. Hahn's marked galley proofs remain in his literary estate, now in the Archives of the Max Planck Society, Berlin (a copy was kindly lent me by R. Neuhaus).

[81] O. R. Frisch's letter to Hahn, Jan. 4, 1939 (from Copenhagen) (Archives of the Max Planck Society, Berlin), edited in Krafft, *Im Schatten der Sensation*, pp. 271 f.

[82] L. Meitner and O. R. Frisch, "Disintegration of Uranium by Neutrons: A New Type of Nuclear Reaction", *Nature,* 143 (1939), 239 f. (communicated Jan. 16, 1939, published Feb. 11, 1939); see also the experimental test by O. R. Frisch, "Physical Evidence for the Division of Heavy Nuclei under Neutron Bombardement", *ibid.*, 276 (communicated Jan. 16, published Feb. 18, 1939). Both are reprinted in Wohlfarth, *40 Jahre Kernspaltung*, pp. 97–100, resp. 101 f. (These results were communicated by Niels Bohr to his American colleagues after he received the famous long telegram with the message by Frisch.) The corresponding experiments by F. Joliot only began when those by Frisch were finished; see Weart's essay in this volume.

[83] Notebook "Chemie III" (see n. 19 above), fols. 18^V f., printed in facsimile by Krafft, *Im Schatten der Sensation*, p. 284, Fig. 30 The measurements began on the evening of Jan. 24, 1939.

84 O. Hahn and F. Strassmann, "Nachweis der Entstehung aktiver Bariumisotope aus Uran und Thorium durch Neutronenbestrahlung; Nachweis weiterer aktiver Bruchstücke bei der Uranspaltung," *Naturwissenschaften*, 27 (1939), 89–95 (communicated Jan. 28, 1939, published Feb. 10, 1939), reprinted in Wohlfarth, *40 Jahre Kernspaltung*, pp. 78–96, and in Krafft, *Im Schatten der Sensation*, 287–293 (facsimile).

85 N. Bohr, "Resonance in Uranium and Thorium Disintegrations and the Phenomenon of Nuclear Fission," *Phys. Rev.*, 55 (1939), 418 f. (communicated Feb. 7, 1939, published mid-March, 1939); then N. Bohr and J. A. Wheeler, "The Mechanism of Fission," *ibid.*, 56, 426–450 (communicated June 28, 1939, published Sept. 1, 1939). The second paper is reprinted in Wohlfarth, *40 Jahre Kernspaltung*, pp. 141–190.

LAWRENCE BADASH

OTTO HAHN, SCIENCE, AND SOCIAL RESPONSIBILITY

Otto Hahn is widely portrayed as a warm, considerate, charming person. The characterization is accurate. In fact, precisely because the personality of this decent human being suffered no great changes throughout his career, he offers us a touchstone to determined the extent of changes in scientists' perceptions of their obligations to society during the twentieth century.

Hahn, as is well known, was a young chemist who anticipated an industrial career. Because his intended employer planned to use him in the international aspects of his business, Hahn left Germany in 1904 to spend several months in England, where his primary goal was to master the language. Since he felt the need for some kind of other activity as well, he secured a place in William Ramsay's laboratory at University College, London. Ramsay, wishing to take advantage of Hahn's background in organic chemistry, and deliberately disregarding his ignorance of radioactivity, asked him to analyze a bowl of radioactive material. So successful was the novice that he discovered a new radioelement, which he called radiothorium. Despite the appraisal by Bertram Boltwood, a noted radiochemist at Yale University, that Hahn's substance was nothing but a compound of "thorium X and stupidity,"[1] Hahn soon convinced the doubters that radiothorium did indeed exist. This turned his future away from industrial organic chemistry into a new direction. In order to learn more about radioactivity before returning to Germany for a now anticipated academic career, Hahn spent the 1905–1906 school year at McGill University, working with that most active of the radioactivity investigators, Ernest Rutherford. Other discoveries followed, for Hahn had almost a King Midas touch for identifying new radioelements, and he played a long and distinguished role as one of the world's leading radiochemists.[2]

By his own admission, Hahn was far more a doer than a thinker. By this is meant more than the distinction between an experimental and a theoretical chemist. Hahn had a reluctance to conceptualize, to look for organizing schemes, to search for deeper meanings, to go a step beyond the evidence. He was quite candid about this when he admitted that he was familiar with the many radioelements possessing extremely "similar" chemical properties but lacked the courage shown by Kasimir Fajans and Frederick Soddy in 1913 when they declared such bodies *identical*, and Soddy named them isotopes.[3]

167

William R. Shea (ed.), Otto Hahn and the Rise of Nuclear Physics, 167–180.
Copyright © *1983 by D. Reidel Publishing Company.*

A quarter of a century later this disinclination toward contemplation was still evident, for when Hahn and Fritz Strassmann discovered the phenomenon of nuclear fission, their statement was so carefully hedged that one might conclude that they themselves knew not what they had wrought: "As chemists we really ought to revise the decay scheme given above and insert the symbols Ba, La, Ce, in place of Ra, Ac, Th. However, as 'nuclear chemists,' working very close to the field of physics, we cannot bring ourselves yet to take such a drastic step which goes against all previous experience in nuclear physics."[4] In fact, it took Hahn's longtime colleague Lise Meitner and Otto Frisch to describe in plain language what Hahn could only hint at: "It seems therefore possible that the uranium nucleus has only small stability of form, and may, after neutron capture, divide itself into two nuclei of roughly equal size . . . . These two nuclei will repel each other and should gain a total kinetic energy of c. 200 Mev . . . ."[5]

Even Hahn's two very readable autobiographies portray a man not given much to introspection, to analysis of behavior of himself or others, to dealing with ideas.[6] I mention these characteristics not to denigrate Hahn in any way, but to show that he was more inclined to workbench activity than to contemplation.

At the interface between science and society the young Hahn also fit this mold. His research in Germany before World War I was facilitated by access to the chemical warehouse of Knöfler and Company, the arrangement being that he would point out to them any discoveries of commercial value. For example, mesothorium I, an isotope of radium, was called "German radium" and marketed at a price well below its more famous relation. More to the point, radioactive thorium products were used in toothpaste; presumably they would cause the teeth to shine, just as traces of radium in paint made it self-luminous. Hahn, who might have questioned this application, is not known to have done so, although by this time it was recognized that the radiations had some harmful effects. As did virtually all his colleagues, he probably reasoned that the amounts of radioactive materials used were so tiny that bodily harm was most unlikely, and he was no doubt aware that radioactive drinking waters, baths, inhalers, and ointments were being peddled across Europe and North America. Few if any users of such preparations were harmed by radiation, because few if any manufacturers sold products of sufficiently high radioactivity. So scientists were spared the discomfort of considering whether they had a social responsibility to a public that might be injured by something with which they were connected. In any case, this was a period in which the prevailing philosophy was "let the buyer beware."[7]

Upon the outbreak of World War I, Hahn, who was a reservist, was mobilized as a sergeant and saw action on the western front during the autumn of 1914. While in Brussels for a rest, in January 1915, he was ordered to call on Fritz Haber, who was in Belgium on a mission for the Ministry of War. Haber was director of the Kaiser-Wilhelm Institut für Physikalische Chemie, in Berlin-Dahlem, and since Hahn headed the Department of Radioactivity Research in the sister Institut für Chemie, it is likely that they were acquainted. Haber explained to him that the war was bogged down in the trenches and that new weapons were needed to break the stalemate.[8] Chemical irritants had earlier been investigated, the object being to make the trenches uninhabitable. Walther Nernst had little success with such products, and Haber's own research on irritants ended in December 1914, when an explosion in his laboratory killed Otto Sackur. By the time of his meeting with Hahn a month later, Haber's institute had embarked in a new direction, toward developing chlorine gas as a weapon, with the goal no longer of moving the enemy's troops but of killing them.[9]

The Hague Convention of 1899 had resolved that "The contracting Powers agree to abstain from the use of projectiles the sole object of which is the diffusion of asphyxiating or deleterious gases." All the major European countries ratified or adhered to this declaration, a fact which was a matter of widespread knowledge.[10] Hahn, therefore, objected to Haber's proposal, but was told that the French had already violated the Convention by use of gas-filled rifle amunition (with little effect). In any case, Haber urged that gas could bring the war to a much more rapid end, ultimately saving lives – a position, with merits and flaws, taken by supporters of gas warfare on both sides in World War I and after, and by those who justified the use of atomic bombs on Hiroshima and Nagasaki.

Hahn, now a lieutenant, and some former colleagues such as James Franck, Gustav Hertz, Wilhelm Westphal, and Erwin Madelung, were transferred to a new unit in Berlin, Pioneer Regiment Number 36, where they received training in the application of poison gas, defense against it, and meteorology. Within a month or two Hahn was back in Flanders, attached to an infantry regiment as their gas pioneer. Once the gas cylinders were installed along the front, an attack was planned on several occasions. But it had to be scheduled 24 hours in advance, to permit bringing up units from the rear, and every time adverse wind conditions forced cancellation.[11] The High Command thereupon decided to move the chlorine cylinders to a line near Ypres, Belgium, where the winds were more favorable. On April 22, 1915, gas from 5000 cylinders along a 3½-mile front was blown into the air, creating a cloud 700 to 1000 yards deep, which drifted into the French positions and broke their resistance. "Nearly an entire division suffocated in agony. Fifteen thousand causalties were counted; one-third were fatal."[12] This event is universally regarded as the beginning of *scientifically* organized chemical warfare.[13]

Hahn and Franck took no part in the operation at Ypres; they were scouting other possible battlefields. By the end of April 1915, Hahn and Gustav Hertz were in Poland, planning support for another offensive. In June they discharged a mixture of chlorine and phosgene in a successful attack. During the subsequent advance Hahn was shocked to observe the agony of the striken Russian troops, caught without gas masks. He recalled: "We tried to use our own respirators to help some of them, to ease their breathing, but they were past saving. I felt profoundly ashamed and perturbed. After all, I shared the guilt for this tragedy."[14]

Until the war's end, Hahn, who in 1916 was transferred to Imperial Headquarters, travelled from one front to another, with periodic tours of duty in Berlin and at the Bayer Chemical Works in Leverkusen. He participated in the evolution of weaponry, from gas cylinders to gas-filled shells, gas mines, gas cannisters, and gas mortars, and personally filled a number of them with cooled liquid phosgene. He helped develop new gases, and he and Franck volunteered to test upon themselves the maximum concentrations of phosgene the gas masks could safely resist. This work was not without danger, and both Hahn and Franck were known to have risked their lives often, in the front lines and in these laboratory tests. Several of their colleagues were killed or seriously injured by gas.[15]

Gas warfare was not congenial to Hahn, but neither were other types of destruction. Moreover, in the face of strenuous daily activity, moral questions receded to the background.[16] In his autobiography he wrote:

As a result of continuous work with these highly toxic substances, our minds were so numbed that we no longer had any scruples about the whole thing. Anyway, our enemies had by now adopted our methods, and as they became increasingly successful in this mode of warfare we were no longer exclusively the aggressors, but found ourselves more and more at the receiving end. Another factor was that we front-line observers rarely saw the direct effects of our weapon. Generally all we knew was that the enemy abandoned the positions that had been bombarded with gas shells.[17]

Morality, in any case, is a delicate flower, strongly affected by "environmental" conditions. Nationalism is one of the most powerful of these modifiers, and we need only refer to examples from the nineteenth and twentieth centuries to establish the point. During the Crimean War, Michael

Faraday was asked his opinion on a scheme to fill old Royal Navy ships with sulfur, ignite them, and let the wind blow them ashore. The object was to force the garrison at Cronstadt to surrender. Faraday — and it would be hard to find a more humane person — did not comment on the morality of the action at all or the use of science for deadly purposes; war was war. He merely claimed a technical inability to predict the result because there had been no analogous test; but he rather suspected that a few gusts of wind would dissipate the sulfur dioxide.[18] A more recent example of nationalism (perhaps it should be called anti-nationalism) is Albert Einstein's willingness in 1939 to urge President Roosevelt to sponsor investigations of uranium fission. Though a pacifist, Einstein recognized that a nuclear weapon in Hitler's hands would be intolerable, and such a threat must be countered.[19]

To return to World War I, there were many who had no scruples, or lost them, about involvement with gas warfare. It was easiest to claim the need to respond to the enemy's behavior, and among the Allies the response was massive. A research and development complex near Washington, at the American University, for example, began with only two buildings when the United States entered the war, but by the Armistice it had over fifty buildings and employed some 1200 scientists. A research team headed by James Bryant Conant devised a process for the manufacture of mustard gas that was vastly superior to the German technique, only to learn that a British group under Sir William Pope had accomplished the same thing shortly before.[20] Flushed with pride in their successes, chemists sought to solidify their position after the war. When the American Chemical Warfare Service stood in danger of dissolution, powerful voices in the scientific community rushed to its defense. They professed the ideals of research for its own sake and of public service, and no doubt were influenced also by economic self-interest. They also carried the effective arguments that their weapons had indeed shortened the war and in future would be needed solely for defensive purposes. Those who addressed the moral issue directly argued that gas was more humane and efficient than bullets or bombs, and they cited persuasive statistics to this effect.[21]

But the public could not be convinced. Posion gas was widely regarded with fear and loathing. It was called barbarous and inhumane; it was felt to be indiscriminate in striking civilians as well as troops; and its behavior was believed to be insidious because some gases could neither be seen nor smelled. In 1922 the United States, the British Empire, France, Italy, and Japan signed a treaty prohibiting "The use in war of asphyxiating, poisonous or other gases, and all analogous liquids, materials or devices," arguing that such use had been "justly condemned by the general opinion of the civilized world."[22]

Otto Hahn could not have been oblivious to these currents of controversy surrounding the use of posion gas. Even during the war his sensibilities must have been shaken when Richard Willstätter only reluctantly acceded to Haber's appeal to work on gas masks,[23] when Max Born bluntly refused to join Haber's enterprise,[24] and especially when Haber's wife in 1915 begged her husband to forsake poison gas as a perversion of science and as a sign of barbarism, and upon his refusal committed suicide.[25] Hahn must equally well have been shocked by the postward condemnation of Haber as inhuman, especially by Allied scientists who themselves worked on gas and whose rage was enflamed when the 1919 Nobel Prize in chemistry was awarded to Haber for his nitrogen fixation process.[26]

Haber, because he was viewed as the instigator of scientific chemical warfare, was the lightning rod for vilification. He himself feared trial as a war criminal and a sentence of death, although no steps were taken against him. Allied chemists could not be blamed for their participation in the same work, since they were merely responding to the enemy's behavior. And, surprisingly, even Haber's German colleagues escaped opprobrium.[27] Haber was reconciled with such pillars of decency as Max Born and Albert Einstein;[28] in 1933 Sir William Pope offered him sanctuary in Cambridge when Hitler's racial policies forced him from Germany;[29] and an overflow crowd defied a Nazi ban to attend a memorial service in Berlin-Dahlem after his death.[30] However, an example of the longevity of contempt associated with his name was the refusal by Ernest Rutherford to shake Haber's hand in 1934, when Born tried to bring the two together. Rutherford was not a man to harbor anger for any length of time, and there is no evidence that his very cordial relations with either Hahn or Pope were interrupted by their poison gas activities, but Rutherford seems to have drawn the line at Haber.[31] Otto Hahn, it would seem, was typical of the vast majority of scientists of his day in willingly participating in the construction and use of weapons during World War I. It is likely, however, that the experience left him more sensitive than most to the question of whether such participation need always be a socially responsible reaction when one's government calls upon a scientist for his services.

The evolution of Hahn's sense of social responsibility, a change in his balance between the inclination to join the crowd and independent thought, apparently progressed further during the Nazi regime. (As indicated earlier, his scientific boldness did not change as much.) It is difficult, if not impossible, to trace in small stages, because his papers and correspondence were destroyed during World War II bombing raids. One step was his leading role in the memorial service to Fritz Haber. Another was his anguish over the many Jews forced from Germany by Hitler's anti-Semitic laws. Lise Meitner, his physicist-colleague for thirty years, was one of these victims. Clearly, science was susceptible to interference by the state. Science was a social activity that could be manipulated or controlled through the government's power of legislation, funding, appointment, and intimidation.

If society could affect science, as was now so obvious, was the reverse possible? Here it would seem that Hahn's experience with poison gas in World War I left no doubts that the answer was yes. Science could certainly affect society. Despite his loyalty to his country, Hahn had no desire to aid Hitler's ambitions. With experience and maturity, the "doer" in Hahn was enriched by a more contemplative side.

The discovery of nuclear fission by Hahn and Strassmann occurred in December 1938; its interpretation by Meitner and Frisch took place in January 1939. Within a short time, Frisch verified an energy release of the order of 200 million electron volts from the splitting of a uranium nucleus, whereas only a few electron volts was the magnitude expected from the most violent chemical reactions.[32] Nuclear reactions sustained by the emisson of neutrons, if they could be carefully controlled, would offer a steady source of heat from which electricity could be generated, as in our present-day reactors. If the reactions were allowed to progress uncontrolled, it might be possible to release so much energy in which a short period of time as to constitute an explosion a million times larger than that attainable by chemical explosives. Such ideas were not whispered about in 1939; they were freely discussed and published.[33] With the outbreak of war, however, secrecy crept more and more over the subject.

In Germany it was not Hahn who brought the possibility of an energy source or a bomb to his government. Georg Joos and Wilhelm Hanle of Göttingen, and Paul Harteck and Wilhelm Groth of Hamburg, fulfilled that function. Nor did Hahn play a leading administrative or technical role in the fragmented and mismanaged German uranium project; chemists, in fact, were apparently scorned by the physicists involved. Although Hahn attended meetings at which progress was discussed, Werner Heisenberg, some other physicists of ability, and a number of Nazi party hacks led that work through its many trials and tribulations. The argument advanced that Heisenberg and the others, as moral Germans, deliberately dragged their feet so as to deprive Hitler of such an awful weapon, is unconvincing.[34] The German project, which failed even to achieve a chain reaction, never got to the point where a decision about constructing a bomb had to be made. It was just an idea trotted out for the military any time the scientists anticipated cuts in funding or personnel.

Hahn spent the war doing what he had done for decades, identifying and studying the characteristics of radioactive and inactive bodies, in this case the fission fragments. This basic research had value to a reactor or bomb project but was not of the scope Hahn was capable of had he embraced such an effort enthusiastically. His research papers, in fact, were allowed to be published, to suggest to Allied Intelligence that no secret uranium work was being pursued in Germany, and because his findings revealed little critical information. In 1944, for example, the Kaiser-Wilhelm Institut für Chemie, of which Hahn was now director, produced a table of the fission products from element number 35 (bromine) to element 59 (praseodymium).[35] It is reported that he quite specifically refused to have anything to do with the military applications of fission.[36]

In 1945, ten prominent German scientists, including Hahn, Heisenberg Max von Laue, and Walther Gerlach, were taken into protective custody by the advancing American and British armies. The Western powers wished to prevent the Russians from commandeering their services, and they also wanted to learn the extent of German scientific accomplishments during World War II. The scientists were taken to Farm Hall, a country estate near Cambridge, England, which had previously been used as an embarkation point for secret agents about to be parachuted into occupied territory. During their several months of internment the conversations of these scientists were secretly recorded, in hopes of obtaining information that might not have been forth-coming during interrogations. Excerpts from the "bugged" conversations have been published, in such books as General Leslie Groves' *Now It Can Be Told* and Samuel Goudsmit's *Alsos*. Groves was the American Army officer in charge of the Manhattan Project, and Goudsmit, a well-known physicst, was the chief American analyst of German scientific achievements. The British, however, are reluctant to admit that they would be so unsporting as to record conversations secretly, and they have steadfastly refused to declassify the transcripts. A recent attempt of mine to see them ended in failure, and numerous other historians have fared no better.[37]

Most pertinent for our consideration are the events of August 1945, when these German "house guests" were informed of the destruction of Hiroshima by a nuclear weapon. They at first shared a sense of disbelief. If German science had not accomplished construction of the atomic bomb, then no other country was capable of it. Surely, it must be a propaganda attempt to get the Japanese to surrender. They gave each other lectures on nuclear physics and on the technical problems to be overcome in order to produce the requisite fissionable material. With more radio and newspaper reports their arrogance changed to bitterness and shame that they had not been the scientists to achieve this technical success. Carl Friedrich von Weizsäcker, a physicist and philosopher, gave his interpretation: "I believe the reason we didn't do it was because all the physicists didn't want to do it, on principle. If we had wanted Germany to win the war we could have succeeded." Otto Hahn replied, "I don't believe that, but I am thankful we didn't succeed." Erich Bagge, a theoretical physicst who was an expert on isotope separation, later remarked, "I think it is absurd for Weizsäcker to say he did not want the thing to succeed. That may be so in his case, but not for all of us." On the morality of the bomb, Karl Wirtz, an experimental physicist, exclaimed, "I'm glad we didn't have it." Weizsäcker felt "it's dreadful of the Americans to have done it. I think it is madness on their part." Heisenberg differed. "One can't say that," he countered. "One could equally well say, 'That's the quickest way of ending the war.'" And this rationale was the only thing that consoled Otto Hahn.[38]

Hahn had been the first informed about Hiroshima on August 6, 1945, by the British officer in charge at Farm Hall. The news completely shattered him, for he felt that his discovery of fission had made construction of the atomic bomb possible, and that he was thus personally responsible for the thousands of deaths in Japan. Long before, he had contemplated suicide, when he first recognized the possible military use of fission; now, with the blame of its realization drawn squarely upon his shoulders, suicide again seemed a way to escape his desolation. Fearing this, Max von Laue remained with him until he passed this personal crisis. Never has social responsibility hit a scientist with such impact.[39]

Hahn learned while still interned at Farm Hall that he was awarded the Nobel Prize in chemistry for the discovery of fission. This added prominence to his already distinguished career, and his wartime anti-Nazi stance made him all the more acceptable to the Allied occupation authorities. Thus, he became a leading figure in the resurrection of German science after the war, an elder statesman who held the confidence of the various factions. In his position as president, he was particularly successful in rebuilding the Kaiser Wilhelm Society, the parent body of the institutes, which was renamed the Max Planck Society.[40]

His wartime recognition of the perversion of science for the construction of weapons and his postwar activity in planning the direction of his country's scientific endeavors now inclined him increasingly toward being a spokesman for social responsibility. In early 1954 he wrote an article on the misuse of atomic energy, which was widely reprinted and also read on the radio in Germany and abroad. The next year he was instrumental in organizing the Mainau Declaration, a statement to the same end which ultimately was signed by over fifty Nobel laureates, and which was issued a week after the similar Bertrand Russell–Albert Einstein Manifesto.[41]

The most publicized and controversial action involving Hahn took place in 1957, when he and seventeen other scientists, who had formed a "Nuclear Physics Group," critizied the policies of the West German government. They had become disturbed by Chancellor Adenauer's drive for rearmament and were quite alarmed at his intention to allow NATO forces to stockpile nuclear weapons on German territory. A letter early in the year to Minister for Defense Franz Josef Strauss, asking him to disclaim any intention by Germany to manufacture or store such weapons, was indignantly rejected. Strauss called their suggestion "outrageous" and said they were hampering his efforts to improve Germany's position relative to the Soviet Union. Germany was not allowed by the Western powers to build nuclear weapons, but he felt that their storage on German soil, and possible tactical use by the *Bundeswehr*, was desirable.[42]

The scientists were unhappy with this encounter and felt the need to appeal to the public. On April 12, 1957, they sent a declaration to the three largest German daily newspapers, pointing out that even tactical nuclear weapons can do enormous damage, especially in the quantity that likely would be used, that the entire West German population could be annihilated by the radioactivity released by the strategic weapons of the day; and that while the nuclear balance of terror had probably preserved the peace for some years, it was unreliable in the long run. "We do not feel competent to make concrete proposals regarding the policy of the Great Powers." But, they continued, "We believe that a small country like the West German Federal Republic can do most for its own defense and for the maintenance of global peace by explicitly and voluntarily renouncing possession of any kind of atomic weapon. In any case, none of the undersigned would be prepared to participate in any way whatsoever in the manufacture, testing, or use of atomic weapons." The names underneath were the elite of scientists residing in Germany, including Max Born, Walther Gerlach, Otto Hahn, Werner Heisenberg, Max von Laue, Josef Mattauch, Friedrich-Adolf Paneth, Fritz Strassmann, Carl Friedrich von Weizsäcker, and Karl Wirtz.

The declaration drew from Chancellor Adenauer both a denunciation in a radio broadcast and an invitation to meet with him and other officials. Their lengthy discussion seems not to have changed anyone's views then, nor is the scientists' effect on later arms control efforts measurable. The appearance of Sputnik shortly thereafter no doubt confirmed each side in its position. But the importance of this event lies in its very public character, in which senior scientists labeled certain scientific activities as socially unacceptable and professed concern that their younger colleagues not be drawn into such work.[43]

These three episodes in the career of Otto Hahn are illustrative of the general pattern of scientific attitudes in the twentieth century toward the application of their knowledge. In World War I he was an important second-rank figure in the development and use of poison gases, and though he found the business distasteful, war was not expected to be pleasant, and his country needed his services. World War II found him opposed to his government's behavior, and while he refrained from research of military value, he was not known for outspoken condemnation of the Nazis. Few could expect to have anti-government behavior tolerated, Max von Laue being a courageous example.[44] Finally, the enormous importance of science in the postwar period led Hahn into public positions which he hoped would influence government, the public, and his colleagues.

More often than not, scientists have felt the call of nationalism in times of war as much as their other countrymen. Even without hostilities, when danger seems apparent, military establishments have little difficulty in staffing their laboratories. The change in attitude toward social responsibility has not led to a corresponding change in the pace of weapons development, but it has affected the number of individuals who take a personal stand to forego participation in such programs. Scientists have grown in political sophistication: governmental behavior no longer goes unquestioned; indeed, it often is vigorously opposed. World War II was a "popular" war in that good and evil were easily discernible, and the Allied citizenry gave overwhelming support to the fight. Scientists also enthusiastically offered their expertise; Sir James Chadwick once mentioned that he knew of only one or two examples in Great Britain of scientists declining weapons work on principle, and Arthur Compton, in his book *Atomic Quest*, described the case of but one young man, Volney Wilson, who asked to be spared from nuclear explosive calculations, yet who, after the Japanese attack on Pearl Harbor, volunteered to resume them.[45] The Cold War, with its testing of nuclear weapons in the atmosphere until 1963, with its science-based missiles, warheads, reconnaissance satellites, and other contributions, and the Vietnam conflict, with its defoliants and electronic battlefield, have been far more controversial than World War II. Disagreement with official policy has given many scientists the incentive to affirm a responsibility to humanity that is higher than allegiance to their country. If the public rarely speaks with a single voice, this too is the case with the scientific community. The important thing is not that scientists may disagree on where their responsibility to society lies, but that they are conscious that a responsibility exists, are vocal about it, and when they speak out they expect to affect policy. Otto Hahn, it would seem, was even more than just an example of this twentieth-century conceptual evolution; he was a leader in the process.

University of California at Santa Barbara

NOTES

[1] Lawrence Badash, ed., *Rutherford and Boltwood, Letters on Radioactivity* (New Haven: Yale University Press, 1969), p. 81. This remark is contained in a letter from Boltwood to Rutherford, dated Sept. 22, 1905, preserved in the Rutherford Collection, Cambridge University Library.

[2] Otto Hahn, *A Scientific Autobiography* (New York: Scribner's, 1966), esp. pp. 11–36. *Otto Hahn, My Life* (London: MacDonald, 1970), esp. pp. 61–74.

[3] Hahn, *Scientific Autobiography*, pp. 50, 99.

[4] Otto Hahn and Fritz Strassmann, "Über den Nachweis und das Verhalten der bei der Bestrahlung des Urans mittels Neutronen entstehenden Erdalkalimetalle," *Naturwissenschaften,* 27 (1939), 11–15. Translation appears in Hans Graetzer and David Anderson, eds., *The Discovery of Nuclear Fission: A Documentary History* (New York: Van Nostrand Reinhold, 1971), p. 47. As Prof. Fritz Krafft has pointed out to me, another reason for Hahn's reticence to conceptualize, at least in this case, was the division of labor that he and Lise Meitner had established over the course of three decades. Interpreting the physics of the phenomenon was her territory and Hahn was not inclined to invade it – nor was he familiar with the data necessary for energy calculations.

[5] L. Meitner and O. R. Frisch, "Disintegration of Uranium by Neutrons: A New Type of Nuclear Reaction," *Nature,* 143 (1939), 239–240.

[6] See n. 2.

[7] See Lawrence Badash, *Radioactivity in America: Growth and Decay of a Science* (Baltimore: Johns Hopkins University Press, 1979), Chs. 9 and 10 for a survey of early medical and commercial uses of radioactive materials.

[8] Hahn, *My Life*, pp. 102–104, 112–118.

[9] Morris Goran, "Fritz Haber," *Dictionary of Scientific Biography*, Vol. V (New York: Scribner's, 1972), p. 622.

[10] *The Hague Declaration (IV, 2) of 1899 Concerning Asphyxiating Gases* (Washington, D. C.: Carnegie Endowment for International Peace, 1915), pamphlet 8, pp. 1–2.

11 Hahn, *My Life*, pp. 118–119, 130.
12 Morris Goran, *The Story of Fritz Haber* (Norman: University of Oklahoma Press, 1967), p. 68. Hahn, *My Life*, p. 119. Richard M. Willstätter, *From My Life* (New York: Benjamin, 1965), p. 265.
13 Mario Sartori, *The War Gases* (New York: Van Nostrand, 1940), p. vii.
14 Hahn, *My Life*, pp. 119–120, 132.
15 *Ibid.*, pp. 121–132.
16 As happened also at Los Alamos in 1945. See Lawrence Badash *et al.* eds., *Reminiscences of Los Alamos, 1943–1945* (Dordrecht, Holland: Reidel, 1980).
17 Hahn, *My Life*, pp. 122–123.
18 L. Pearce Williams, *Michael Faraday* (New York: Basic Books, 1965), pp. 482–483; Williams, ed., *The Selected Correspondence of Michael Faraday* (Cambridge: Cambridge University Press, 1971), Vol. II, pp. 749–751, 767. I am indebted to Prof. Williams for this information.
19 Letter from Einstein to Roosevelt, Aug. 2, 1939. Printed in Morton Grodzins and Eugene Rabinowitch, eds., *The Atomic Age* (New York: Basic Books, 1963), pp. 11–12.
20 Gilbert Whittemore, "World War I, Poison Gas Research, and the Ideals of American Chemists," *Soc. Stud. Sci.*, 5 (1975), 135–163, esp. pp. 150–151. W. J. Pope, "Chemistry in the National Service," *J. Chem. Soc. Lond.*, 115 (1919), 397–407, esp. pp. 400–402. Goran, *Haber*, p. 169.
21 Whittemore, "World War I," pp. 136, 155, 158, 161. Duncan C. Walton, *The Medical Aspects of Chemical Warfare* (Baltimore: Williams & Wilkins, 1925).
22 Warren G. Harding, "Address of the President of the United States submitting the treaties and resolutions approved and adopted by the Conference on the Limitation of Armament," 67th Congress, 2nd sess., Senate doc. No. 125, Feb. 3, 1922 (Washington, D.C.: Government Printing Office, 1922).
23 Hahn, *My Life*, p. 121.
24 *The Born-Einstein Letters* (New York: Walker, 1971), p. 20.
25 Goran, *Haber*, pp. 71–72.
26 M. Goran, "Haber." "Award of the Nobel Prize to Professor Haber," *Science*, 51 (1920), 208–209. For information about nationalism among scientists in World War I, see Lawrence Badash, "British and American Views of the German Menace in World War I," *Notes Rec. Roy. Soc. Lond.*, 34 (1979), 91–121.
27 Hahn, *My Life*, p. 131.
28 *Born-Einstein Letters*, pp. 19–20.
29 Goran, *Haber*, p. 165.
30 *Ibid.*, pp. 173–174. Hahn, *Scientific Autobiography*, pp. 107–113.
31 Max Born, *Physics in My Generation* (London/New York: Pergamon, 1956), p. 223.
32 O. R. Frisch, "Physical Evidence for the Division of Heavy Nuclei under Neutron Bombardment," *Nature*, 143 (1939), 276.
33 For a survey of nearly 100 papers on fission published in 1939, see Louis A. Turner, "Nuclear Fission," *Rev. Mod. Phys.*, 12 (1940), 1–29, esp. pp. 20–21.
34 Robert Jungk, *Brighter Than a Thousand Suns* (London: Gollancz and Hart-Davis, 1958; Pelican edition, 1964), pp. 97–102. For examples of criticisms of Jungk's thesis, see reviews by E. U. Condon, *Science*, 128 (1958), 1619–1620; C. P. Snow, *New Republic*, 139 (1958), 18–19; Geoffrey Barraclough, *Spectator* (1958), 843–844.
35 Hahn, *Scientific Autobiography*, pp. 169–172; *My Life*, p. 173.

[36] O. R. Frisch, "Scientist who Opened the Way to Atom Bomb," *Observer* Aug. 4, 1968. Hahn, *My Life*, p. 185.

[37] R. V. Jones, *Most Secret War* (London: Hamish Hamilton, 1978), pp. 481–483. Samuel Goudsmit, letter to the author, Oct. 28, 1977. Edward J. Reese, Military Archievs Division, U.S. National Archives, letter to the author, June 2, 1978. Margaret Gowing, former chief historian of the British Atomic Energy Authority, letter to the author [Apr. 1978].

[38] L. R. Groves, *Now It Can Be Told* (New York: Harpers, 1962), pp. 334–336. Samuel Goudsmit, *Alsos* (New York: Schuman, 1947), pp. 134–139. David Irving, *The Virus House* (London: Kimber, 1967), pp. 11–16. Jones, *Most Secret War*, pp. 481–483.

[39] Hahn, *My Life*, p. 170. Groves, *Now It Can Be Told*, p. 333. Irving, *Virus House*, pp. 16–17. Jones, *Most Secret War*, pp. 481–483. While not explicit, Hahn's words in the reference cited here suggest that he meant suicide. The authors who have quoted the Farm Hall transcripts, also cited here, have adopted the same meaning. There is, however, good reason to believe that this interpretation may be incorrect. Fritz Krafft has kindly informed me of a letter from Hahn to Robert Jungk, Nov. 26, 1956, preserved in the Archives of the Max Planck Gesellschaft, in which Hahn denies having thought of suicide. Also, Ernst Berninger has generously provided me with a photocopy of Hahn's Farm Hall diary, which, while showing his deep depression, indicates that Hahn was in control of himself. The entry for Aug. 6, 1945, contains the following lines: "I almost lose my nerve again a little at the thought of the new, huge misery." After discussion with his interned colleagues, "I soon go to my room to go to bed. But Heisenberg comes to talk to me. He was obviously sent by the Major. But I really don't need any consolation" . . . because he had not taken part in weapons development. This entry has been published in Dr. Berninger's *Otto Hahn in Selbstzeugnissen und Bilddokumenten* (Reinbek bei Hamburg: Rowohlt, 1974), pp. 89–90.

[40] Hahn, *My Life*, pp. 175, 178.

[41] *Ibid*., pp. 219–221. For Mainau, on July 15, 1955, see *N.Y. Times* July 16, 1955, 3:1. For Russell-Einstein, on July 9, 1955, *ibid*., July 10, 1955, 1:2 and 24:4.

[42] Hahn, *My Life*, pp. 221–222.

[43] *Ibid*., pp. 222–226. Note that Born and Paneth were among a number of scientists who left Germany in the 1930s and returned after the war. "18 German Physicists Ban Work on Nuclear Weapons," *N.Y. Times* Apr. 13, 1957. "German Physicists Protest Nuclear Weapons," *Science*, **125** (1957), 876.

[44] Alan Beyerchen, *Scientists Under Hitler* (New Haven: Yale University Press, 1977), pp. 65–66.

[45] A. H. Compton, *Atomic Quest* (London: Oxford University Press, 1956), pp. 41–42. J. Chadwick, interview with the author, Feb. 11, 1970.

NEIL CAMERON

THE POLITICS OF BRITISH SCIENCE
IN THE MUNICH ERA

On October 14, 1937, Lord Rutherford became ill, the result of sudden complications from a minor hernia that he had had for years. Hospitalization and surgery followed, but after a brief rally, his condition began to deteriorate rapidly. On October 19, the most famous of all Cavendish directors was dead. He had shown almost no previous indications of frail health: he had been working in the Cavendish as usual, and had been planning a trip to India for the following year.[1]

The sudden news of his death brought shock and grief to leading scientists of many countries, especially those like Niels Bohr and Otto Hahn, who were former students. But the British scientific community was overwhelmed. J. G. Crowther, science correspondent of the *Manchester Guardian* and one of the best-known Marxist writers on science in the 1930s, later recalled that when he heard the news, he wept more than he did when his own father died.[2] Such reactions were typical. J. J. Thomson, a man not normally given to hyperbole, commented: "His death just on the eve of his having [newer and more powerful research equipment] is, I think, one of the greatest tragedies in the history of science."[3]

For Sir Frank Smith, the head of the Department of Scientific and Industrial Research, the shock was even greater. Rutherford had chaired the advisory committee of independent scientists that served the DSIR, and Smith had worked closely with him for years. He was so shaken that he experienced a temporary breakdown and left his position — which brought Edward Appelton, one of Rutherford's ablest progeny, into control.[4] Smith was only a few years younger than Rutherford, and the intensity of his reaction can perhaps be explained partially by the sudden reminder of his own mortality. But even for the other late Victorians who still held so many prominent positions in British science, it was more than the passing of the most brilliant and successful of their contemporaries that disturbed them. Sir Henry Dale, the Director of the National Institute for Medical Research from 1914 to 1942 and president of the Royal Society during World War II, touched on a broader issue when he later lamented that "Rutherford was lost to us at a time when we and all the world most surely needed the strength of his wisdom and the guiding power of his genius."[5]

William R. Shea (ed.), Otto Hahn and the Rise of Nuclear Physics, 181–199.
Copyright © 1983 by D. Reidel Publishing Company.

At first glance, Dale may appear to be indulging in a pardonable exaggeration. It would not be disputed, of course, that Rutherford had been able to boast of an unparalleled record of achievement in experimental physics, from the early work at McGill to the great days of his Cavendish team in the 1930s. Nor is there any compelling reason to reject the immense number of testimonials from contemporaries and former students to the attractiveness of his strong and simple character.[6]

But Dale's tribute is not so much one to a great Cavendish director as it is to Rutherford as sage and statesman — as a lost *political* leader of British and world science. Yet this was, in fact, what Rutherford had become by the 1930s, although it was not a rôle with which he was entirely comfortable. To understand how this had come about, and the effect that his death had on the relations between science and politics, it is necessary to review briefly some more general developments that had taken place in British culture and society since the turn of the century.

At the end of the nineteenth century the worlds of science and politics were still linked in the persons of aristocratic amateurs. Lord Salisbury, even while Conservative Party leader and Prime Minister, maintained his own laboratory at Hatfield. Other aristocrats and men of independent means might confine a larger proportion of their energies to science exclusively, but they could still move in the same circles as the social and political leadership of the country from an early age. Rayleigh, for example, who was Cavendish director after James Clerk Maxwell, was married to Arthur Balfour's sister and was a personal friend of Salisbury.[7]

On the surface, little had changed by the early 1900s. Balfour and Richard Burdon Haldane kept alive the amateur tradition, cultivating philosophy, science, and politics at the same time. Scientific research still depended to a very considerable extent on private patronage: Rayleigh could afford to donate his Nobel Prize for an extension to the Cavendish, while Arthur Schuster could retire early from his physics chair at Manchester and provide a handsome endowment to help lure Rutherford from McGill. Such methods could give adequate support for the most brilliant representatives of the rising generation, but occupational opportunities in physics and several other areas of science were still limited.

But the rising power of new industrial wealth, the growth of political democracy, and the attendant development of specialization and professionalization were all beginning to effect major changes in both science and politics by the time Rutherford returned permanently to England in 1907. The electoral methods and party organization required by democracy, for example, tended to make politics itself a more "professional" activity. Men of aristocratic origins could sometimes make the transition successfully, but if they were ambitious to attain the highest offices, they increasingly found it necessary to adopt the single-mindedness of their middle-class competitors. Thus Churchill, who belonged to the same generation as Rutherford, kept something of the versatility of the aristocratic style, but he also had to conform to many of the political practices brought in by men like Joseph Chamberlain and Lloyd George.[8]

At the same time, more and more of the scientific community was coming to be made up of university-based professionals who depended on science for their livelihood, aided by the growth in demand for teachers of service courses for medicine and engineering. The prestige conferred by Nobel Prizes, Copley Medals, and similar awards, and the power provided by the research directorships, created a new group of scientific leaders, men like Thomson and Rutherford in physics, F. G. Hopkins in biochemistry, and A. V. Hill and Sir Henry Dale in physiology and medicine. As he outpaced all rivals, Rutherford increasingly became a sort of symbolic chieftain of this group.

But increasing middle-class hegemony in both politics and science involved more than a simultaneous and parallel change in social composition; it also brought about a subtle change in the kind of social contacts that were carried on between members of the two groups. In politics, to begin with, a somewhat more heterogeneous political élite was bound by its very nature to limit the effectiveness of the polymath-amateur, whose ideal political environment was one composed of men of broadly similar cultural background and disposition, who could integrate at least some portion of the polymath's philosophic and scientific ideas into their own view of politics.

This ideal had little chance of success in the Lloyd George—Baldwin—MacDonald era. Balfour and Haldane sought a middle path between Baconian positivism (which Sir Richard Gregory's *Nature* was making into the official rhetoric of professional science) and traditional politics (which made scarcely any allowance for the importance of science and technology). Their attempts to bring about some kind of synthesis of philosophy, science, and politics were not very successful and were sometimes a source of derision among their colleagues (this was especially the case with Haldane). But these projects arose out of a reasonable conviction: that some kind of bridge ought to be maintained between the world of abstract thought and the world of ordinary political action.

The professional politicians were still willing enough to use the services of the polymath-amateurs in particular offices of state, but only as possessors of special knowledge or as adornments. Ideas had become simply the tools of the trade; the business of government absorbed all the real energy and interest. Lengthy examination of abstract moral principles, metaphysical presuppositions, or the more general implications of science could at best provide material for political platforms, at worst be a tiresome nuisance. Hence the main importance of Balfour and Haldane did not come from any success at promulgating the ideal of scientifically informed politics, but simply from the fact that they frequently held important offices which gave them the opportunity to promote scientific interests.

The last years of both men's lives were filled with examples of the changing of the guard, in both politics and science. Balfour, for example, was Prime Minister in the early 1900s, then served on a level with men like Edward Carson and Bonar Law in World War I, then spent the 1920s as Lord President of the Council (the minister responsible for science). At the beginning of the decade, when J. J. Thomson was about to give up the presidency of the Royal Society, he offered it to Balfour; the latter politely declined, on the grounds that it should go to a professional scientist – as it has ever since. Thomson, at the same time, was relinquishing the Cavendish to Rutherford and becoming the master of Trinity College; he was the first professional scientist to head Trinity, the first master not to have taken holy orders, and the probable beneficiary of Balfour's influence with Lloyd George.[10]

Both Haldane and Balfour died, with a timing that could scarcely have been more appropriate if they had planned it, at the end of the decade. Rutherford was by then president of the Royal Society. In giving the eulogy for Balfour, he concluded that "His place will not easily be filled."[11] This is the conventional stuff of farewell tributes, but in this case it was an understatement. The new social and political order that was gradually emerging in England could find *no* substitute for Balfour.

It was true, of course, that many of the functions that had once been filled by the polymath-amateurs could now be split up between the senior scientific civil servants and scientist-administrators, like Sir Frank Smith and Sir Henry Tizard. But even the one or two men of outstanding ability were not so placed as to be able to deal effectively with all the problems that would be of mutual concern to scientists and men in public life. For one thing, the government scientific bodies still occupied very junior and uninfluential positions in the overall bureaucratic system. The scientific civil servants were often charged with the task of minimizing expenditure by setting up more effective combinations and consolidations of existing machinery. Their institutions lacked the status, the vitality, and the scale of operation to enter prominently into the political calculations of the House or the Cabinet.

Most of the scientist-administrators were men of cautious temper and diffident manner, and these tendencies were reinforced by their recognition of the limited power that they controlled. Nor could they assume that the latter was only a temporary weakness resulting from the newness of bodies like the Department of Scientific and Industrial Research and the Research Councils. It arose at least as much from the sharp division that had developed between the theory and practice of British politics and the theory and practice of British science.

Politics was changing not only as a result of professionalism, but also of collectivism. A general movement toward the positive state had begun even in the late Victorian era; the turn of the century had brought fierce attacks on the very intellectual and moral foundations of liberalism, launched from such different sources as reactionary cultural critics and militant trade unionists. The World War had brought about at least temporary major ventures in governmental direction of capital and labor, and the postwar weaknesses of the very industries which had been the mainstays of nineteenth-century prosperity drew more and more politicians into accepting at least some degree of interventionism and "planning" (a magic word by the 1930s). Conservative paternalism, Liberal humanitarianism and reformism, and Labourite socialism all made their contribution to a much wider enthusiasm for increasing state power than could be found at the turn of the century.

But science, the social institution which most spectacularly symbolized change and progress, was also the one which had the least difficulty in continuing to maintain the intellectual and organizational conventions of the Victorian age. While these were also retained in many other areas of English society, they were generally on the defensive. Individualism, aristocratic style, honorific reward, academic detachment, and voluntary association were all under attack, but not often in the Royal Society's meeting rooms or in Rutherford's Cavendish. The University laboratories had remained the centers of scientific life, and they could point to a triumphant record of continuing success; their leading spokesmen remained confident, optimistic, and self-assured.

From about 1935, a small but energetic group of radical scientists and scientific publicists, entranced by the example of the Soviet Union, began crusading for the general adoption of a more Marxist-cum-Baconian model of scientific research and organization.[12] But they made only a limited number of converts among the younger generation, and the biochemist Hopkins was their only major supporter among the older men, who still held the dominant positions in the laboratories and associations.[13] Baconian and collectivist models of science had little influence with the general public as well, partly because both the quality and mass-circulation press had helped to make outstanding individual scientists into public heroes, whose glory had been only a little sullied by the "Chemists' War" of 1914–1918. In particular, the popular cult of Einstein that had started up in the 1920s, while it was viewed with amusement or even resentment by some British physicists, increased the general fascination with science and the prestige of other scientific heroes.[14]

What all this meant was that there was an almost exact correspondence between a scientist's prestige and the amount of power he possessed, not only in his own laboratory or professional association, but in influential political circles and in society at large.[15] And it was Rutherford who held the commanding position in this hierarchy of prestige.

Perhaps because Rutherford's whole life has something of the appearance of an unimpeded triumphal procession, little note has been taken of the fact that the status he had achieved by the 1930s was of a quite unusual kind, even for a very famous scientist, and it had important political consequences.[16] He was still a major research physicist, the brilliant achievements of the younger Cavendish men in the 1930s still tended to reflect more glory on him, and he was also an elder of the tribe.

Most scientists of his generation or older experienced an inverse correlation between their accumulation of honorific duties and administrative powers and the amount of prestige and influence remaining to them in the centers of research activity. In a scientific environment that was becoming increasingly professionalized and competitive, this division between older and younger men tended to be that much more noticeable. Senior men frequently compensated for reductions in their "vertical" influence by increasing their "horizontal" commitments, by assuming more duties in the general scientific bodies and in various university and government councils.

Rutherford had joined the ranks of these senior statesmen of science, but he had done so while still remaining the central presence in the most famous and active physics laboratory in the world. He joined the very small group of men whose continued creativity made it possible for them to hold prestige and influence in the eyes of several scientific generations. But unlike the rest of these, he was also seen as the supreme practitioner of the most prestigious branch of scientific enquiry, a man who was compared by his colleagues with Faraday and even Newton. Even though most of the important discoveries in his field were being made by his students, he was still very much their guide and inspiration, not merely their formal administrator. The two forms of prestige were mutually reinforcing. Those who were thought of as representing mature judgment and wisdom in the scientific community were put in the shade by Rutherford's position of leadership in physics, especially if they had become "extinct volcanoes" in their own fields. Those who were forging ahead in research, on the other hand, bowed to his age and experience. And Rutherford's self-assurance and blunt but amiable manner made it possible for him to make the most of this combination of advantages.

Ramsay MacDonald did not have many occasions for important contacts with science and scientists; when he did, he was likely to turn to Richard Gregory, an old friend, for aid. But Baldwin, himself a product of Trinity College, found it natural to turn to Rutherford, and they, too, became good friends. By the early 1930s, Rutherford had achieved an unprecedented and unchallenged level of authority in the social system of British science.[17] He could not be described as a dictator, save perhaps in his determination to maintain the deep but narrow focus of Cavendish physics, and he was not authoritarian in his fundamental philosophical and political convictions. But in his wider relations with the scientific community, he had assumed something of the character of a feudal monarch. His kingdom contained other powerful nobles, with their own loyal retinues and complex systems of reciprocal obligation and homage, but none disputed who was the greatest in the land.

At the time of his death, there was, however, no crown prince; not even Chadwick could be so described. There were plenty of examples of brilliant achievement by the rising generation in physics. Men like E. V. Appleton and P. M. S. Blackett were also beginning to take on a wide range of public and administrative duties. But no single one of the younger men had forged so far ahead of his rivals as Rutherford had done earlier in the century. In fact, it has been suggested that the choice of W. L. Bragg, an X-ray crystallographer, to succeed Rutherford at the Cavendish was precisely because it was so difficult to choose between the claims of the nuclear physicists.[18] Rutherford's long hegemony had made it more difficult for all of them to stand out clearly as authoritative figures in their own right.

It followed that when Rutherford died, there was no general agreement among either scientists or laymen as to who would now be expected to act as the spokesman or spokesmen for British science — and in a period when political issues could scarcely be ignored. The scientists who could make the strongest claims to do so were not necessarily drawn from physics but were distributed fairly evenly among the major disciplines. The most influential were usually those who, like Rutherford, combined a long-sustained record of major achievement in their field with extensive social contacts built up through the laboratories, professional bodies, universities, and governmental advisory committees. Most had served on the Council of the Royal Society, and several also served on the executives of the less prestigious general associations. One or two came from wealthy or socially prominent families, or had close personal connections with politicians, important intellectuals, or higher civil servants. It would obviously be difficult to provide a definitive list of these individuals, but it can safely be assumed that any such list should include the names of Frederick Gowland Hopkins (biochemistry), A. V. Hill (physiology), J. B. S. Haldane (genetics), Henry Dale (medicine), and Alfred Egerton (chemistry). The most prominent of the experimental physicists could be added, with Sir William Bragg as the senior figure, as well as the theoreticians, C. G. Darwin, R. H. Fowler, and G. I. Taylor.

To these would also have to be added Frederick Lindemann, the head of the Clarendon Laboratory at Oxford, and Henry Tizard, Rector of Imperial College. Lindemann's close personal friendship with Churchill and his wide range of upper-class social contacts gave him considerable political influence, even though he had made plenty of enemies at Oxford, and his two attempts to run for the university as an Independent Conservative had both ended in failure. Tizard had negligible influence in the world of politics, but his long service on bodies like the Department of Scientific and Industrial Research, the Aeronautical Research Committee, and the Council of the Royal Society had provided him with many friends in science, industry, and the Service Ministries. Lindemann and Tizard had already become personal enemies, as a consequence of their disagreements on Tizard's Air Ministry Aerial Defence Committee; something more will be said of this famous clash later.

Whether in making public political pronouncements or in advocating policies for science behind the scenes, all of these men suffered from the disadvantage of having only limited recognition and support from some sectors of the political nation: those who commanded respect from Cambridge dons did not necessarily do so from Conservative ministers, and so on. In the absence of a central unifying figure, the three main executive officers of the Royal Society — the president and the two secretaries — assumed a considerable importance in defining some sort of response to the issues raised by the deteriorating international climate. The nature of that response can be partly explained by the tradition they had inherited from Rutherford, and partly by their own personal temperaments.

Rutherford had a notorious dislike for the kind of speculative interpretations of science associated with James Jeans and Arthur Eddington, but he had his own ways of influencing the British scientific community in their choice of general assumptions. Physicists and scientists in other disciplines have frequently commented on this influence, which was felt even by strong and sometimes resistant personalities like Chadwick and Blackett. As the latter put it:

[His] single-minded and passionate interest in the nucleus led him to sometimes decry the importance and interest of other branches of physics and still more so of other sciences. Though this depreciation was more jocular than serious, his prestige was such that even a joke from Rutherford's mouth was apt to become a dogma in lesser men's minds. No very young physicist could be totally unaffected by his famous crack: "All science is either physics or stamp collecting", or by the often implied assumption that it only needed some further progress in physics to allow us to deduce from first and physical principles the facts and laws of lesser sciences ... Rutherford undoubtedly should have been sympathetic to Compte's [sic] hierarchy of sciences with physics placed right at the top, and with each science deducible in principle from the next one higher up the list.[19]

This may be a slight overstatement, but Blackett's general impression has been supported by many other scientists.[20] Rutherford helped maintain a positivist tinge in British scientific ideas, but he also did a considerable amount to ensure that it would remain no more than a tinge. He could be as hostile to utilitarianism as he was to Eddingtonian idealism. And his joyful celebration of the most abstract and general of the sciences gave little encouragement to those who would make a general "scientific way of thinking" the basis for all thought and political action.

Rutherford thus had a very considerable effect on the general political style of British science, without doing much to enjoin any particular party opinions. Most of the large number of scientists who have produced tributes to Rutherford scarcely thought it necessary to mention his political ideas. However, Mark Oliphant, who had many political arguments with him, describes him as "what would be called nowadays, a woolly liberal," suspicious of both communism and extreme conservatism, fond of Baldwin.[21] He was also, throughout his life, a strong supporter of the British Empire, as well as being one of its most celebrated advertisements. But his real influence was not on behalf of liberalism or imperial sentiment: it was rather in the compelling authority with which he assigned science and scientists their place in the general scheme of things.

In both background and outlook he represented the narrowly focused professionalism that had been emerging in science since the late nineteenth century. He had been quite willing to accept the idea that he was entitled to only a limited, if significant, degree of influence in areas outside science, however great his eminence in the eyes of his colleageus and the general public. He had gradually accepted an increasing burden of public and governmental responsibilities, almost as additional honors comparable to those heaped on him by the scientific community. In dealing with political authorities he had demonstrated moderation, common sense, tact, and some skill, and he had done so with the general approval of his colleagues.[22] But while he enjoyed the company of public figures, he was never really at home in the world of politics; he was a physicist above all.

He probably increased his standing, and that of scientists in general, with men like Baldwin and Hankey by his very unwillingness to make many pronouncements on divisive political issues; the public celebrity of figures like Einstein and Russell was usually in almost inverse proportion to their influence on governments. On the other hand, his approach was one which assumed implicitly that scientists could or should have no part in actually changing a general climate of opinion. For the most part, he thought and acted within the intellectual and institutional framework of Victorian liberalism.

In general, Rutherford tended to treat the violence, political fanaticism, and war fever of the 1930s as perverse and largely incomprehensible aberrations, comparable to natural cataclysms or outbreaks of epidemic disease. His public responses were invariably civilized and humane, but he appears only gradually to have come to recognize the full implications of Hitler's accession to power in Germany, for example. In April of 1933, he had still, in a letter to a German colleague, been able to refer to "the anti-semitic troubles" as a "strange effervescence" which he hoped would in no way affect his friend.[23] By the following month he had come to realize that persecution of German scholars of Jewish origin was taking place on a national scale, and he became a major figure in the Academic Assistance Council which gave help to émigrés, first as a member of the executive and later as president. But it is worth noting that this decision on his part owed a great deal to the skilful lobbying of William Beveridge on a Cambridge weekend, when G. M. Trevelyan and Gowland Hopkins were also present.[24] It was not that Rutherford had any reservations, but his anger and indignation had to be fired.

Once it was, what he wanted to do was provide practical help; he hoped to avoid "politics and recriminations."[25] He continued with this cautious approach. At the mass meeting to support the Academic Assistance Council that was held in the Albert Hall in October of 1933, he had been notified that Austen Chamberlain was anxious to avoid any direct hostile references to Germany.[26] Chamberlain need scarcely have worried; in the event, his own speech was considerably more fiery than Rutherford's. The latter talked of avoiding any "petty spirit of national or sectional hostility" or of "political antagonism."[27] He tried to follow this procedure as much as possible, later writing in *The Times* that the Council had been formed "in the conviction that the universities form a kingdom of their own," and linking the persecution and expulsion of the Jews to a more general world refugee problem.[28]

In other words, while he strongly opposed Nazi racial policies, he tried as much as possible to act as if the maintenance of good political relations between England and Germany could be treated as an entirely separate issue, rather as if the Nazis had controlled only a social movement rather than the machinery of the state. In 1934, when alarm was growing in England about the vulnerability of the country to aerial attack, he could still express the hope in a public speech that aerial warfare could be abolished by international agreement.[29] It was in the same year that Lindemann, whose mind worked on very different lines, wrote his famous letter to *The Times* advocating that "the whole weight and influence of the Government should be thrown into the scale" to see if scientists could develop a counter-weapon to the bomber.[30]

It is not surprising that Baldwin and Rutherford liked each other: both men could be tough enough, but preferred the methods of quiet diplomacy and peaceful compromise whenever possible. Thus, what Rutherford left behind at his death was not so much an identifiable political philosophy as a style and an attitude, one that had appealed to many Englishmen both inside and outside science.

Of the Royal Society executives who found themselves acting as political spokesmen for British science after Rutherford's death, two had a very similar approach, one did not. Sir William Bragg, the distinguished X-ray crystallographer who was president of the Royal Sociaty from 1935 to 1942, was even more cautious in international relations than Rutherford, and more hopeful that some kind of reconciliation or understanding with Germany was possible. Sir Alfred Egerton, the chemist who was physical secretary, was less sanguine than Bragg, but he had little inclination to launch dramatic or controversial initiatives of any kind in public affairs. A. V. Hill, the biological secretary, was, however, a man of rather different temper.

Bragg, born almost a full decade before Rutherford, was a gentle and kindly Victorian, deeply Christian in his beliefs.[31] His daughter and biographer has described him as one of the "most unpolitical of persons," and he declined an offer to stand for Parliament for a Combined Universities seat in 1937.[32] He had lost a son in World War I and hated the prospect of another War. When he had been approached with the suggestion that he take an executive position on the Academic Assistance Council, he had written to Rutherford that he was afraid the organization would do more harm than good by angering the new German government.[33]

Bragg's origins had been comparatively humble; Egerton came from quite a different world.[34] He could trace his descent from a lord chancellor of James I and had aristocratic relations on both sides of his family. He was an Etonian, married to the sister of Stafford Cripps, and had spent seventeen years of his life at the Clarendon, working with Lindemann and Thomas Merton, Thomas Merton, another gifted scientists of independent means. All three of these men could afford to cultivate a variety of interests, but only Lindemann included a fascination with politics and power among them. Egerton was widely travelled and a talented artist; he had political convictions which could be described as "liberal-conservative" or "conservative-liberal," but few burning ambitions. He had considerable personal charm and was on good terms with almost everyone in the scientific community.

Bragg, while once a scholarship student at Cambridge, had spent much of his life in Australia and was really as much identified with the Royal Institution as with any university. Egerton was mostly associated with University and Imperial College and with Oxford. But A. V. Hill, although he did much of his teaching in London, was very much a Cambridge don.[35] He had begun his career in mathematics and physics at Trinity, but his interests had later shifted to physiology, in which he had won a Nobel Prize. He was married to the sister of John Maynard Keynes, but he was not particularly sympathetic to the kind of high-minded academic liberalism often associated with Cambridge.

Hill was a close personal friend of Egerton's, and got on well with Bragg, but he had a far greater appetite for controversy than they did. He was not really all that different from Egerton in convictions: he, too, could probably be best described as a liberal-conservative or conservative-liberal. But while he had no sympathy for the Marxism of J. B. S. Haldane, and had become a bitter opponent of Lindemann after crossing swords with him on the Tizard Committee, he was somewhat like both the Communist and the High Tory in at least one respect: he tried to actively influence the general climate of opinion in politics.

Hill's preference for sharply worded and pungent arguments comes out throughout his correspondence with other scientists in the 1930s and 1940s, but it can be seen to best advantage in the Huxley Memorial Lecture that he gave in 1933.[36] The lecture was an eloquent defense of the ideals of independent scholarship and political liberty. Hill began it by arguing that all civilized societies had granted certain immunities and privileges to men of science, not because the latter always acted wisely as individuals, but because they contributed to a body of knowledge that was the common property of all sane human beings. He attacked both the Soviet idea of science as "the handmaiden of social and economic policy" and the attempts, endorsed by some scientists, to apply a crude social Darwinism to human affairs.

He defended the Second Charter of the Royal Society, which required that body to avoid interference in religious, philosophical, and political questions. He conceded that the prohibition was difficult to observe in the 1930s, but he was unimpressed by most of the excursions that had been made by scientists, and others, into areas outside their own expertise. He made fun of Bernard Shaw's science, Frederick Soddy's economics, and Einstein's project for putting young mathematicians in lighthouses: "one pities the poor sailors who would depend upon their lights!"

Hill believed that the growth of internationalism, cooperation, and tolerance had been one of the greatest achievements of the European intellectual community, but he was willing to declare boldly that totalitarian politics might make it necessary to fight for those values. He took some pains to argue that attachment to international understanding did not presuppose the abandonment of a reasoned patriotism, and that those who hated war most could still be among the best fighters.

In the mid-1930s Hill joined his friend Tizard on the latter's Committee for the Scientific Study of Air Defence, an Air Ministry advisory committee that had got its start in the hope that its members could come up with a "death ray," but had eventually wound up guiding the development and operational adoption of radar. The creation of this committee had coincided with Baldwin's creation of a more unwieldy but more politically powerful body, the Air Defence Research Committee that reported directly to the Committee of Imperial Defence. The ADRC had come about through the urgings of Churchill and Lindemann, and the latter had somewhat reluctantly joined Tizard's Air Ministry Committee, which was supposed to report to both the Air Ministry and the ADRC. This arrangement might have led to difficulties even apart from the controversy that broke out between Lindemann and the other scientists about the value of research on aerial mines. Hill and Blackett eventually resigned from the Tizard Committee in 1936, so that Lord Swinton, the Air Minister, was forced to reconstitute it without Lindemann.[37] Swinton and Tizard had recognized that the bitter disagreement with someone as powerful as Churchill's closest scientific advisor virtually compelled them to give a "political" weight to what had been a "technical" committee. Not only was Lindemann replaced by Edward Appleton, but Rutherford had been added as a "senior consultant."

When the Munich crisis came in 1938, Churchill pressured the government into bringing Lindemann on to the ADRC. Tizard, who was furious, commented bitterly to Hill that "It is rather clever the way they have waited until Rutherford was well dead, so to speak, before pressing this again." [38] He bowed to the pressure, but was successful in demanding that Hill be added to the ADRC as well. All of these experiences had convinced Hill that something had to be done to improve central scientific coordination with the Cabinet, especially as the threat of war became greater every month. In the meantime, he was now an active member of both committees, hard at work on aerial defense.

Bragg's hopes for peace and reconciliation were greater. His own response to the Munich crisis had taken the form of a broadcast on the BBC of what was practically a sermon; in fact, it was his use of the expression "moral rearmament" that gave Dr. Frank Buchman the idea of adopting it for his Oxford Group, although this did not make Bragg happy.[39]

Hill's energetic efforts to guarantee that British science would be prepared for war was thus yoked rather oddly with Bragg's continuing hopes that peace could be preserved. It was Bragg's determination to keep open the lines of communication with Germany that led to one of the most remarkable international scientific meetings of the 1930s. It was to coincide with preparations of quite a different kind.

In September 1938, Bragg attended an International Documentation Conference in Oxford and was there approached by the head of the Prussian State Library in Berlin. As a result of their discussions, he announced in his Royal Society presidential address in December, he had received a letter from the president of the Kaiser Wilhelm Gesellschaft suggesting that the latter body undertake some common enterprise with the Royal Society, and further suggesting an exchange visit.[40] A delegation was duly prepared, composed of F. G. Donnan and A. J. Clark; John Dover Wilson, the distinguished Shakespearean scholar, was also added. Ironically, their visit to Germany took place in March 1939, the month that Hitler took over the rest of Czechoslovakia and that the British gave their guarantee to Poland; they could hardly have been very comfortable.

By then the British were wondering how they should deal with the Germans at international conferences in general. Egerton noted, in the diary that he kept of Royal Society business, that it was "rather awkward now that Hitler has broken his word. We should show disapproval, but let it still be felt that we can be hospitable towards true representatives of science even from those countries which we may consider are politically misguided."[41] The following day (March 21), he noted that he had met with Sir Henry Lyons (director of the Science Museum) and Sir Albert Seward (vice-chancellor of Cambridge and foreign secretary of the Royal Society), as well as Hill: "Seward agreed that in spite of political difficulties that it would be best to treat with the Kaiser Wilhem Gesellschaft as if nothing had happened. . . ".[42]

It then transpired that Otto Hahn would be one of the four Germans who would be coming in June. The British were, of course, by then quite familiar wih the famous nuclear fission experiments that Hahn had carried out in December, and Egerton noted in May that he expected an "interesting discussion" on fission in the meeting they were arranging for Hahn at the Royal Institution.[43]

But this was not all that Egerton had on his mind. Early in June, Hill came up with a new scheme for the coordination of scientific knowledge in Britain that he thought would be necessary in the event of war. It amounted to getting the Cabinet to accept the idea that he, Egerton, and Bragg, and perhaps one or two other top scientists, should form a Scientific Advisory Committee to serve the Committee of Imperial Defence. Tizard had tried proposing a somewhat similar scheme about a year previously, but his own suggestion, which had not been accepted, had been for a committee made up of the leading representatives of the government scientific establishments. However, when Hill and Egerton approached him, he gave his blessing to Hill's proposal.[44]

The formal contact took the form of a letter from Bragg to Chatfield (the new Minister for the Co-ordination of Defence), which was written a week before the Germans arrived.[45] From this time on, Bragg was to find all of his energies taken up with pushing the claims of this advisory committee. At about the same time, the Royal Society executives set about compiling a national register of scientific talent in Britain, to be used by the authorities when war came.[46]

When Hahn came, he pleased his listeners by saying that "German scientists think of Ernest Rutherford as probably the greatest figure in science that the world has produced since. Isaac Newton."[47] Hahn presented a paper on atomic fission, and while there is no published record of the entire discussion that followed, Egerton noted the main points:

Bohr led the discussion and was in very good form. Then we had tea and resumed the discussion with the President, who had returned from the banquet, now in the chair. Oliphant, Thomson, and B. from Cockcroft's Laboratory spoke, Bohr also. A good part of the discussion ranged over the possibility of making the splitting self cumulative. Bohr pointed out that it could not be dangerously so and that the process intrinsically possessed a safety value as it were. The discussion was a great success.[48]

His one regret about the soirée held on the same day was that not enough distinguished people from outside science put in an appearance.[49]

Bohr, if still mistaken about the possibility of an atomic bomb, had been more disturbed by what he took to be the general outlook of his British colleagues. About a year after the gathering, Max Born wrote to Einstein that at the time that it had taken place, Bohr was "quite excited and shocked by the indifference of most of the British people to the imminent danger of war, and he tried to convince all people he met . . . about the danger . . . though at that time many people here . . . still believed in appeasement."[50]

Perhaps many people did, but Bohr, like many others, probably gained an exaggerated idea of the extent of British complacency from the fact that the earnest gestures of men like Bragg were highly visible; the determined preparations of men like Hill and Tizard were not. But it should be noted that even the latter, following in the tradition of Rutherford, were trying to strike a balance between the claims of national self-defense and resistance to tyranny, and those of international cooperation and friendship.

The combination of military and diplomatic initiatives taken by the Royal Society executives can thus be seen as a characteristic reflection of a more general ambivalence in the British public mind in the 1930s; the culmination of Bragg's efforts in the Hahn meeting is also a reminder that even the most brilliant scientists can do no more than speculate on the course of future events. Niels Bohr was to find himself as mistaken about the possibility of making an atomic bomb as Bragg had been about the prospects for Anglo-German reconciliation. In science, as in politics, the "appeasement era" presented more difficult and uncertain alternatives than has yet been realized in the popular imagination.

McGill University

NOTES

[1] For Rutherford's illness and death, see A. S. Eve, *Rutherford* (Cambridge: Cambridge University Press, 1939), pp. 424–427.

[2] J. G. Crowther, *Fifty Years With Science* (London: Barrie and Jenkins, 1970), p. 189.

[3] J. J. Thomson, "Lord Rutherford: An Obituary," in Eric Homberger, William Janeway, and Simon Schama, (eds.), *The Cambridge Mind* (London: Cape, 1970), pp. 177–181 (p. 181). This is the obituary Thomson originally prepared for both *Nature* and the *Cambridge Review* in 1937.

[4] Charles Goodeve, "Frank Edward Smith, 1876–1970," *Bio Mem. FRS,* **18** (1972), 525–548; Ronald Clark, *Sir Edward Appleton* (London: Pergamon, 1971), pp. 93–120.

[5] Henry Dale, "Some Personal Memories of Lord Rutherford," in his *An Autumn Gleaning* (London: Pergamon, 1954), pp. 165–182 (p. 182). On Dale, see W. S. Feldberg, "Henry Hallett Dale, 1875–1968," *Biog. Mem. FRS,* **16** (1970), 77–175. Like Rutherford, he had been a scholarship student at Trinity College, but as an undergraduate; he had entered a year before Rutherford arrived as a research student.

[6] On Rutherford's character and example, see, e.g., the commemorative collection edited by J. B. Birks, *Rutherford at Manchester* (London: Heywood, 1962). There are an immense number of comparable tributes. While many of these were occasioned by the academic habit of using hallowed names and events as justifications for social and professional gatherings where bitterly critical accounts would be out of place, they can hardly be dismissed. The practice of treating Rutherford as something of a national monument was already long established before he died and explains much of the nature of his importance outside the world of science.

[7] Robert Bruce Lindsay, *Men of Physics – Lord Rayleigh* (Oxford: Peramon, 1970).

[8] W. L. Guttsman, "The Changing Structure of the British Political Elite, 1886–1935," *Brit. J. Socio.,* **2** (1951), 122–134.

[9] (Fourth) Lord Rayleigh, *The Life of Sir J. J. Thomson* (Cambridge: Cambridge University Press, 1942), p. 204. See also the same author's *Lord Balfour in his Relation to Science* (Cambridge: Cambridge University Press, 1930). Balfour served on the Council of the Royal Society in both 1907–1908 and 1912–1914, when his political responsibilities were considerable.

[10] Lloyd George to Rayleigh, Jan, 17, 1941, quoted in the latter's life of Thomson, p. 205.

[11] Ernest Rutherford, "Presidential Address," *Proc. Roy. Soc. Lond.,* **A 130** (1931), 239–259 (p. 240).

[12] On the radical scientists, see Gary Werskey, *The Visible College: A Collective Biography of British Scientists and Socialists of the 1930s* (London: Allen Lane, 1979); Neal Wood, *Communism and British Intellectuals* (New York: Columbia University Press, 1959), pp. 121–151.

[13] Sir Richard Gregory, like his friend H. G. Wells, gave general support to the Baconian arguments for "socially planned" science, so the radicals had the support of the editorial columns of *Nature*. But the Victorian positivism of the older men sat uneasily with the Marxism of younger men like John Desmond Bernal. Gregory had pushed his own views for two decades, without making any noticeable impression on men like Thomson or Rutherford. This point deserves laboring, because the immense number of books and articles churned out by a few radicals (notably J. B. S. Haldane and Lancelot Hogben)

has often caused laymen to exaggerate the general extent of radical ideas among British scientists in the 1930s. But the Association of Scientific Workers, the radical-led union, even though it took in technicians as well as scientists, was actually only about the same size as the Royal Society for most of the decade.

[14] For the lukewarm response of British physicists to the postwar excitement about Einstein's general theory of relativity, see Ronald Clark, *Einstein* (New York: World, 1971), pp. 300–302.

[15] Such a close correlation between prestige and power cannot always be taken for granted. In World War II some of the most distinguished figures of the 1930s had less influence than men building new careers and reputations in the government establishments.

[16] The specialized literature on the social and political relations of British science between the wars often concentrates on radical movements and professional organizations; see, e.g., Roy and Kay MacLeod, "The Social Relations of Science and Technology, 1914–1939," in *The Fontana Economic History of Europe*, Vol. V: *The Twentieth Century* (London: Collins-Fontana, 1976), pp. 301–363; Russell Moseley, "Tadpoles and Frogs: Some Aspects of the Professionalization of British Physics, 1870–1939," *Soc. Stud. Sci.*, 7 (1977), 423–446. Little attention is given to Rutherford's part in shaping the whole system. This is Hamlet without the Prince.

[17] Bernal's *Social Function of Science* (London: Routledge, 1939) contains several criticisms of British scientific organization that could imply personal criticism of Rutherford and the "gerontocracy" in leading positions, but neither he nor any of the other radicals directly challenged any of Rutherford's public pronouncements.

[18] J. G. Crowther, *The Cavendish Laboratory 1874–1974* (New York: Science History, 1974), pp. 269–270.

[19] P. M. S. Blackett, "Memories of Rutherford," in Birks, *Rutherford at Manchester*, pp. 102–113 (p. 108).

[20] See, e.g., the other contributions to Birks's *Rutherford at Manchester*.

[21] Mark Oliphant, "The Two Ernests [Lawrence and Rutherford]," *Phys. Today*, 19, No. 9 (1966), 35–49 (p. 43).

[22] Rutherford's carefully worded reply to Johannes Stark's complaining letter about criticism of the Nazis appearing in *Nature* was made after consultation with other leading British scientists; see Eve, *Rutherford*, p. 380.

[23] Rutherford to Georg von Hevesy, Apr. 3, 1933, reprinted in Eve, *Rutherford*, pp. 370–371.

[24] William Beveridge, *Power and Influence* (New York: Beechhurst, 1955), pp. 235–236.

[25] *Ibid.*, p. 236.

[26] Walter Adams (General Secretary of the Academic Assistance Council) to Rutherford, Oct. 1, 1933, Rutherford Papers, Cambridge University Library, A File.

[27] Ernest Rutherford, Albert Hall Address of Oct. 3, 1933, quoted in Eve, *Rutherford*, p. 375.

[28] Ernest Rutherford, "The Wandering Scholar," *Times* (May 3, 1934).

[29] Ernest Rutherford, Address to the Cambridge University Branch of the Democratic Front [undated, 1934], quoted in Eve, *Rutherford*, p. 389.

[30] F. A. Lindemann, letter to the *Times* (Aug. 8, 1934).

[31] On the older Bragg, see E. N. da C. Andrade, "William Henry Bragg, 1862–1942,"

Obit. Not. Roy. Soc. Lond., 4 (1943), 277–300; G. M. Caroe [his daughter], *William Henry Bragg* (Cambridge: Cambridge University Press, 1978).

32 Caroe, *Bragg*, p. 110.

33 Bragg to Rutherford, May 4, 1933, Rutherford Papers, Bragg File.

34 Lady Egerton's *Sir Alfred Egerton* (London, 1963) is a memoir with papers, but is privately printed and hard to come by. See also D. M. Newitt, "Alfred Charles Glynn Egerton, 1886–1959," *Biog. Mem. FRS*, 6 (1960), 39–64.

35 Bernard Katz, "Archibald Vivian Hill, 1886–1977," *Biog. Mem. FRS*, 24 (1978), 71–149.

36 A. V. Hill, "The International Status and Obligations of Science," Huxley Memorial Lecture of Oct. 1933, reprinted in his *Ethical Dilemma of Science* (New York: Rockefeller Institute Press, 1960), pp. 205–221.

37 Blackett to Lord Swinton, Jul. 15, 1936; Hill to Swinton, same date; Public Record Office Papers, CAB 16–132. The best account of the dispute with Lindemann is in Ronald Clark, *Tizard* (Cambridge, Mass.: MIT Press, 1965), pp. 105–148.

38 Tizard to Hill, Nov. 24, 1938, Hill Papers, Churchill College, Cambridge, File 2/5.

39 Caroe, *Bragg*, pp. 110–115. Bragg's continuing hope that reconciliation with Germany was possible was perhaps more characteristic of the older scientists. J. J. Thomson's comment on the Munich crisis was that Chamberlain was "magnificent," and his biographer, another scientist of the older generation, agreed; Rayleigh, *Thomson*, p. 286.

40 Caroe, *Bragg*, p. 115.

41 A. C. G. Egerton, Diary Entry of Mar. 20, 1939, "Personal Notes Relatings to Meetings and Business of the Royal Society, 1938–," Egerton Papers, Royal Society Archives.

42 Egerton, Diary Entry of Mar. 21, 1939, Egerton Papers.

43 Egerton, Diary Entry of May 11, 1939, Egerton Papers.

44 Egerton, Diary Entry of June 1939 [exact date not given], Egerton Papers.

45 Bragg to Lord Chatfield, Jul. 13, 1939, Public Record Office Papers, CAB 27–712.

46 Ernest Brown, Minister of Labour, to Bragg, Dec. 9, 1938, Hill Papers, File 2/1; "The Royal Society and the Central Register," *Notes Rec. Roy. Soc. Lond.*, 2 (1939), 176–178.

47 Quoted by Sir Henry Dale, *Autumn Gleaning*, p. 182.

48 Egerton, Diary Entry of June 23, 1939, Egerton Papers.

49 *Ibid.*

50 Born to Einstein, Apr. 10, 1940, *The Born-Einstein Letters* (London: Macmillan, 1971), pp. 138–140.

THADDEUS J. TRENN

WHY HAHN'S RADIOTHORIUM SURPRISED
RUTHERFORD IN MONTREAL

In the autumn of 1904, Hahn went to London to spend some time in the laboratory of Sir William Ramsay, an organic chemist of great repute who had strong ties to Germany. His fame was only that much more increased by his recent discovery of most of the inert gases, for which he received the Nobel Prize in 1904. Ramsay's main interest had already begun to shift to radioactivity. One question open at the time was that of the atomic weight of radium. Marie Curie had obtained a value of about 225 using chemical techniques.[1] The spectroscopic results of Carl Runge and Julius Precht, however, seemed to indicate that radium was even heavier than uranium.

Ramsay had the idea that if a small quantity of radium could be isolated in the manner done by Friedrich Giesel and then allowed to combine with organic compounds, perhaps the atomic weight of radium could be deduced from a determination of the molecular weights of these compounds. It was the ideal project for the young organic chemist Hahn. Ramsay provided him with a dish containing nearly 20 grams of barium carbonate with the instructions that he was first to extract the several milligrams of radium as radium bromide using Giesel's method.[2] This method required less stages of fractionation, making use of the bromide, than did Curie's method which utilized the chloride. Ramsay expected to begin the actual research with organic compounds of radium once this fractionation exercise was completed. But this research program took an unexpected turn.

As Hahn began the process of fractional crystallization, obtaining an everincreasing concentration of radium in the less soluble fraction, he observed that the more soluble fraction exhibited a radioactivity of its own not characteristic of radium. Particularly striking was that it yielded a radioactive gas, emanation, identical with that given off by *thorium* compounds. The emanation from thorium could be distinguished from that of radium by measuring the time it took for the activity to decay. Thorium emanation decays to halfvalue in about 1 minute, while radium emanation requires about 4 days. Hahn's radiosubstance was highly active compared with ordinary thorium, which in contrast appeared to be inactive. Ernest Rutherford and Frederick Soddy had already discovered what they had construed as the active constituent of thorium; but this constituent, thorium X, characteristically decayed

201

William R. Shea (ed.), Otto Hahn and the Rise of Nuclear Physics, 201–212.
Copyright © 1983 *by D. Reidel Publishing Company.*

to half-value in about 4 days. Hahn's radiosubstance indicated no noticeable reduction in activity over time. But since it yielded the emanation characteristic of the thorium series, Hahn correctly concluded that it must be a member of that genetic series. He called it radiothorium, because it seemed to be the radioactive constituent of thorium itself. Radiothorium − not thorium − appeared to be the real parent of the active constituent, thorium X, discovered by Rutherford and Soddy. And since radiothorium was so closely associated with thorium, Hahn suggested that it might be the "first slow transformation product of thorium."[3]

Now, Hahn was essentially correct about radiothorium being a transformation product of thorium, one genetically prior to thorium X. But this was really not the *first* slow transformation product. To help us keep track of developments, it will be useful to refer to Figure 26, which I have prepared as an aid to follow what was actually going on. The first slow transformation product of thorium is mesothorium I, which takes about 6 years to decay to half-value, compared with only about 2 years for radiothorium. Thus, mesothorium is the first slow transformation product in the genetic sense, and it decays about three times slower than radiothorium. It is this relatively long-lived mesothorium, also discovered by Hahn a few years later, that proved to be the real origin of Hahn's radiothorium. It turns out that mesothorium I and radium have the same chemical properties, for they are not different elements but are isotopes. Thus, mesothorium I is really radium with an atomic weight of 228 in contrast to the main variety of radium with atomic weight 226. The most significant difference between them is their origin. Radium-228 is a member of the thorium series and is specifically the first daughter product of the main variety of thorium, thorium-232, by slow alpha decay. Radium-226 is a member of the uranium series and the direct descendant of the rather long-lived intermediate radiosubstance ionium, chemically identical to thorium.

It was 1911, however, before it became clear that methods of chemical separation could not isolate the isotopic varieties of radium from each other or those of thorium from each other. The sample which Ramsay had provided Hahn contained not only radium-226 but also some radium-228, the relative amounts depending to some extent upon the uranium/thorium ratio of the original mineral deposit. In this particular case, the original ore was thorianite from Ceylon, which contains a high percentage of thorium in relation to uranium. Radium-228 can be compared to a cuckoo egg. At first it is not apparently different from radium-226, but in the course of time, this radium-228, mesothorium I, begins to contaminate the pure genetic uranium series

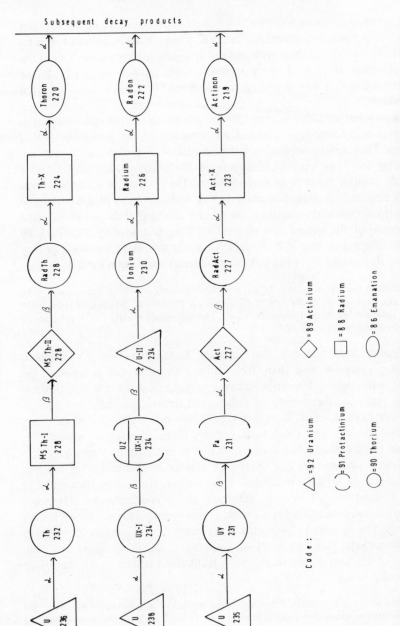

Fig. 26. The three radioactive series geometrically presented as an aid in following displacements in the chemical character of the daughter products.

with radiothorium and with the subsequent decay products of the thorium series. Once the radiothorium is "hatched" from the mesothorium I, it is no longer chemically identical with radium but now is identical with thorium. Radiothorium is thus as easily separable from radium chemically as the cuckoo is from its pseudo-siblings in the nest. This is what Hahn achieved in London.

Ramsay was enthusiastic about this result, which seemed even more interesting than the originally planned confirmation of the atomic weight of radium. They quickly drafted a letter to *Nature* in the spring of 1905 announcing this "new element."[4] It was still customary, incidentally, to refer to such isotopic varieties as new elements. The reaction of the experts was highly skeptical. Bertram Boltwood at Yale was something of a confidant of Rutherford in matters chemical. On April 9 Rutherford dashed off his first impression of the alleged new element of Hahn and Ramsay: "I'll back 10 dollars," he wrote, that it is thorium X.[5] Early that May, Boltwood in turn handed Rutherford down the following judgment upon Hahn's result:

I am positive that the substance he got was thorium-X mixed with a little radium. If he made any experiments of the rate of decay he was thrown off the track by the activity of the radium rising, as the thorium-X fell off. I have obtained exactly similar precipitates in working with other minerals.[6]

Presumably Boltwood felt that thorium X had somehow contaminated Hahn's preparation, and that Hahn had merely succeeded in isolating the contaminant. Again, it was not known by scientists at the time that thorium X was really another variety of radium and hence chemically not directly separable from it. Thus, Boltwood's edict was not implausible at the time.

Hahn was well aware that he lacked any formal training in radioactivity, but his unexpected finding encouraged him to write Rutherford saying that he wished to pursue this line of research in Montreal. Rutherford agreed that he could come in the late autumn if he wished. Rutherford then sailed to New Zealand for the summer, returning about mid-September. Meantime Boltwood had conducted some chemical experiments designed to test Hahn's findings. The judgment he reported to Rutherford on September 22 was even less sympathetic than before: "I am confident still that it is only a new compound of thorium X and stupidity."[7] Rutherford replied on September 28 in a similar vein:

In regard to the compound of "thorium-X and stupidity" — very neatly put. You may be interested to know that Hahn is coming over to work with me this year. I think he wants to know how to measure things radioactive so as to be able to prove his elements.

In fact, I think he will "*take physic* and *throw up* his element." He arrives in a few days.[8]

Hahn arrived on schedule about October 1, blissfully unaware that a strong current of expert opinion was running against his findings. But he had the facts at hand and the courage of his own convictions. He also had specimens of radiothorium and other radiosubstances, brought for purposes of demonstration and further research. It was but a matter of days before the opposition completely crumbled in the face of the evidence. The first to go was Boltwood, who had meanwhile been able to "pump" a German colleague of Hahn's from University College, London, Leo Guttmann. On October 4, Boltwood wrote Rutherford:

From what I could get out of him, . . . there seems to be more reason than I had supposed for assuming that Hahn may really have found something. You will have undoubtedly sized up the situation by this time, and I should be extremely interested to learn your conclusions. I suppose that it is not impossible that there may be an intermediate radioactive product between thorium and thorium-X, and that thorium itself . . . may undergo a rayless change and be itself non-radio-active. If the intermediate product had a fairly long average life, then this would explain Hahn's results.[9]

Rutherford had meantime been able to pump Hahn himself, and he shared his observations with Boltwood on October 10:

Hahn has arrived and settled to work and seems a keen man . . . with not much physical knowledge but the latter I hope to rectify. I think from what he has shown me there is no doubt that he has separated a very active fairly permanent constituent from thorium. His specimens are possibly not quite so active as he thought but are certainly over 20,000. . . . He thinks that he has separated a constituent between thorium and thorium-X pointing to the conclusion that thorium itself . . . is non-active.[10]

Ordinary thorium is not really inactive but alpha decays so weakly that it was long thought to be rayless. Hahn's radiothorium is indeed an intermediate radioactive product and the immediate parent of thorium X.

Confronted with the facts, Rutherford conceded the point to Hahn. But to realize what this admission seemed to cost at the time, it is necessary to return to the research program which Rutherford and Soddy had completed just a few years earlier. They had searched in vain for the "inseparable constituent" of thorium and could not obtain thorium free of radioactivity. They accordingly attributed the residual activity and the power of producing their thorium X directly to thorium as a specific property of that element. If thorium was known to be an element, it was not at all clear at the time what status Hahn's radiothorium might have. Although such radiosubstances were still called new elements, it was evident early on that there was no room in the periodic table for all of them, and that they behaved quite differently than ordinary chemical elements. But Rutherford and Soddy had thought that they were dealing with the transformation of known elements into something else by the process of atomic disintegration. Had they based their case upon the behavior of a mere *constituent* after all? And did the atomic transformation proceed without disintegration or less of mass in some cases? Rutherford's nebulous reference to Hahn's radiosubstance as an "active constituent" harks back to the difficulties he had shared with Soddy on these very questions. If Hahn was right, as certainly appeared to be the case, then at least some of the issues they had laid to rest had to be considered anew.

As early as 1901 Rutherford and Soddy had experimental evidence which suggested that the production of thorium emanation was directly due to thorium as a specific property of that element. They were on the verge of claiming this to be the first case of transmutation to be observed in nature, where one element (thorium) produced another (an inert gas), when they suddenly withdrew this suggestion just before the end of that year.[11] The reason for their hesitation was new experimental evidence establishing the existence of thorium X, an "active constituent" of thorium compounds, which could be separated from thorium chemically and which produced the emanation.[12] Rutherford and Soddy therefore abandoned the position that the activity and the production of the emanation were specific properties of an element. Thorium X was an unknown substance as likely to be an impurity as anything else, and they accordingly wisely abandoned their prima facie case for transmutation.[13]

But if the activity of thorium were due to an active constituent, it was not unreasonable to expect that they should have been able to obtain thorium free of this source of activity. No matter how hard they tried, however, they found it impossible to obtain inactive thorium. About 25 percent of the total activity remained with the thorium. As an explanation they initially suggested that "perhaps there was yet another active consituent more closely resembling thorium in chemical properties and hence not as easy to separate."[14] This suggestion is remarkable, for in effect they *were* dealing with a second "active constituent," namely radiothorium, inseparable because its chemical properties are identical with those of thorium.[15] But a proper understanding of this fact still lay ten years in the future. An important revision of their theoretical account in mid-1902 dispensed with active constituents altogether, and the residual activity of thorium was reassociated with the transformation of the thorium atom itself.[16] Not only did this simplify the entire theory, but it also meant that the atoms of one element, thorium, were transforming themselves into a product, thorium X, with distinct chemical characteristics. Natural transmutation thus clearly involved well-known chemical elements.

Hahn's findings placed this secure basis in jeopardy, for it meant that thorium was possibly inactive after all. Did the various transformations, then, concern only unweighable amounts of unknown active constituents associated with thorium compounds? As has been mentioned, there was not sufficient space in the periodic table to accommodate all such radiosubstances, and furthermore it was not at all clear at the time whether these were even related to the elements as traditionally construed. Rutherford and Soddy designated them "metabolons," to distinguish them from ordinary elements.[17] Hahn's radiothorium seemed to imply that the theory of Rutherford and Soddy might not concern ordinary chemical elements but just these metabolons. While their theory could still account for radioactivity, of course, such an implication would tend to isolate the phenomenon of radioactivity from the domain of traditional chemical elements. In one sense, of course, radioactive change is remote from the chemical domain, since it involves only the nuclei of the chemical atoms and not their shell electrons. But in both cases, it is the atoms of the familiar chemical elements which are involved, and in this sense radioactivity is by no means a peripheral phenomenon. While not chemical in kind, the transformations and the radiations do involve true chemical elements. But matters were by no means so clear in 1905, and hence Hahn's radiothorium could have misled one to believe that the theory of Rutherford and Soddy concerned only a peripheral phenomenon.

It is not possible to determine just to what extent Rutherford was aware of all that Hahn's radiothorium could have implied for him at that time. And thus the degree to which this really played a role in Rutherford's particularly strong aversion to radiothorium before his meeting with Hahn is equally uncertain. There is no doubt that one reason for his disbelief was simply a general suspicion of any result coming from Ramsay's laboratory. While this was most likely partly the reason, it is insufficient to account for Rutherford's initial prejudice and subsequent reaction once confronted with Hahn's evidence. Rutherford was no chemist, but he was quite aware of the distinction that he had formulated with Soddy between metabolons and elements. If nothing else, he may have been nagged by a deep-seated feeling that Hahn's result tended to cut radioactivity from its moorings in the domain of ordinary elements. Certainly he was worried about that second "active constituent" which Soddy and he had found to be "inseparable" and which they finally dismissed as unnecessary, attributing the residual activity directly to thorium. If Hahn had succeeded in separating this particular "active constituent," then why had Soddy and he failed? Surely this question must have been very much in his mind. Fortunately, Rutherford himself provides us with several valuable clues to his own thinking at the time.

He had just given a series of eleven lectures at Yale University that March, and these were published in 1906 under the title *Radioactive Transformations*. In the chapter specifically dealing with changes in thorium, Rutherford provided a detailed account of the work he had done with Soddy. There seemed to be no question that elemental thorium was itself radioactive. He summarized this key section as follows:

The results, so far obtained, are completely explained on the disintegration theory put forward by Rutherford and Soddy. On this theory, a minute constant fraction of the atoms of thorium becomes unstable every second, and [each] breaks up with the expulsion of an alpha particle. The residue of the atom after the loss of an alpha particle becomes an atom of a new substance, thorium-X.[18]

That was in the spring of 1905, at a time when Hahn's results had just been announced. The preface of Rutherford's book is dated June 1906, and he points out to the reader that he has "thought it desirable to incorporate the results of the many important investigations which have been made since the lectures were delivered."[19] There is an insert of nearly two pages on "Radio-thorium," portions of which now follow with commentary:

There has been a considerable difference of opinion as to whether thorium is a true radioactive element or not, ie. as to whether the activity of thorium is due to thorium itself, or to some active substance normally always associated with it. Some experimenters state that by special methods they have obtained an almost inactive substance giving the chemical tests of thorium. Some recent work of Hahn is of especial importance in this connection.[20]

Now Rutherford and Soddy had "proven" that the residual activity associated with thorium was a specific property of the thorium atom and not due to some second "active constituent." This stood in contrast to the repeated claim of Hofmann and Zerban that thorium is really inactive.

Returning to the insert, in the next portion, Rutherford provides the physical details which converted him to accept Hahn's findings:

Working with the Ceylon mineral, thorianite, which consists mainly of thorium and 12 per cent of uranium, Hahn was able by special chemical methods to separate a small amount of a substance comparable in activity with radium. This substance, which has been named 'radiothorium', gave off the thorium emanation to such an intense degree that the presence of the emanation could be easily seen by the luminosity produced on a zinc sulphide screen. Thorium-X could be separated from it in the same way as from thorium, while the excited activity produced by the emanation decayed with the period of 11 hours characteristic for [this stage of the] thorium [series]. The activity of radiothorium seems to be fairly permanent. . . ,[21]

But this raised the question of the origin of radiothorium. Just because it was associated with thorium did not necessarily mean that it was *produced* by thorium. After the discovery and confirmation of isotopes a few years later, the truth of this observation became even clearer. Ionium is also an isotope of thorium and hence "associated" with it, but ionium is not a product of thorium. For Rutherford there were just two possibilities: "there can be little doubt that radiothorium is either the active constituent mixed with thorium, or, what is more probable, that it is a product of thorium."[22] Only the latter alternative could maintain the bond between radioactive phenomena and the element thorium, and Rutherford clearly preferred this option. "It seems probable," he wrote, "that this active substance is in reality a lineal product of thorium intermediate between thorium and thorium-X."[23] Whether its association with thorium was genetic or otherwise, a new dimension had been added to the vexed issue of "whether the activity of thorium is due to thorium itself, or to some active substance normally always associated with it."[24] A genetic association, with radiothorium as a product of thorium, certainly kept this element within the domain of radioactive phenomena. But transferring the residual activity from thorium to a *product* of thorium suggested that elemental thorium might not be truly radioactive. The expulsion of rays was often considered the hallmark of radioactivity, and for Rutherford this was especially true. On the question of the relation between radioactive change and rays he once said, typically, that the expulsion of the rays *is* the change.[25] Yet the insert continues:

The results of Hahn suggest that the transformation of thorium itself may be rayless, but that the succeeding product, radiothorium, gives out rays. Further results are required before such a conclusion can be considered as definitely established, but the results so far obtained by Hahn are of the greatest interest and importance.[26]

Thus, one way or the other, thorium seemed to have become less "radioactive" than before. If the activity was due to a chance constituent, then elemental thorium might not even undergo radioactive change. By opting for radiothorium as a *product* of such a radioactive change, Rutherford could at least preserve the genetic connection. But even this was apparently insufficient to guarantee that thorium itself was truly radioactive. The problem was complicated by the fact that Rutherford was not quite sure just how to reconcile the disintegration theory of radioactivity with this sort of "rayless" change.[27] Since for him the expulsion of alpha and beta particles constituted radioactive change, it was difficult to account for radioactive change without such expulsion. Although he hoped that such changes really involved atomic disintegration where the particle is expelled "at too low a velocity to ionize the gas," Rutherford alternatively suggested that such a change may consist in only "a rearrangement of the parts constituting the atom without the projection of a part of its mass."[28] On the evidence available to Rutherford at the time, there was no way to dismiss this latter possibility, which stood as a threat to the generalized disintegration theory of radioactivity. Thus Rutherford had very good reasons for doubting Hahn's claim in the beginning and equally good reasons for treating it with respect once he was converted. It is this complex of circumstances which contributed to Rutherford's surprise.

But there was a happy ending. Hahn stayed on for nine months, until the summer of 1906. He succeeded in showing that the activity of his radiothorium decreased with a half-life of about two years. This line of research, done in competition with Boltwood, led him later to the discovery of mesothorium as the origin of radiothorium. And from a comparison of the radiations from several thorium preparations, Hahn was able to show in 1907 that thorium itself does give out a true alpha radiation of very low energy.[29] Any suspicions that may have survived concerning the status of thorium as a true radioactive element were thus finally laid to rest. It was as Rutherford believed: the alpha particles expelled from thorium were of very low energy but they were there all the same. The generalized disintegration theory of radioactivity had withstood yet another test unscathed.

Hahn always said that it was just luck that led him to discover radiothorium and prove this fact to Rutherford. Soddy and Rutherford had failed to separate what they referred to as the inseparable active constituent for a very good reason. They had been working with thorium compounds, and radiothorium is isotopic with thorium. But Hahn was, in effect, trying to separate the radium from residual thorium products — a relatively easy task. Others have credited Hahn with a special "nose" for new radiosubstances, and evidence tends to support this opinion. While at McGill, Hahn also found the short-lived radiosubstance thorium C in the latter stages of the thorium series. Even more remarkable, however, he gave a repeat performance of his London success, this time with actinium and right in Montreal. One of Rutherford's previous research students, Tadeusz Godlewski, had carefully examined the actinium series, reporting in 1905 that actinium yielded a very active substance analogous to thorium X. He observed that actinium X, as Godlewski designated his substance, in turn generated the gaseous emanation of actinium. Actinium itself appeared to be inactive. Hahn, however, found and announced in April 1906 an alpha-active intermediate product between actinium and actinium X. This intermediate radioactinium, as Hahn called it — another isotopic variety of thorium — was the direct parent of Godlewski's actinium X. Godlewski must have unknowingly separated this product from his actinium, as Rutherford pointed out, for otherwise the actinium would have emitted alpha rays.[30]

Once again Hahn had succeeded in picking out a radiosubstance right from under the nose of another researcher. Others had been in the field ahead of him, but Hahn was gleaning "gold." He went on to discover many other radiosubstances, including both kinds of mesothorium, several branching decay products in the thorium and the radium series, the main variety of the isotopes of protactinium, and the isomer uranium Z.

We have seen how Hahn began his career in radioactivity. The initiation rites were performed by the highest experts, and Hahn was found acceptable. He had unwittingly challenged the lion in his own den, and then came in to tame him. It is not difficult to imagine what would have happened had Hahn been any less of the lion tamer — *dompteur* — than he was.

Max-Planck-Institut für Physik und Astrophysik

NOTES

[1] Frederick Soddy, "Radioactivity," in Thaddeus J. Trenn, ed., *Radioactivity and Atomic Theory* (London: Taylor & Francis, 1975), pp. 58–59.

[2] O. Hahn, "Über ein neues, die Emanation des Thoriums gebendes radioaktives Element," *Jahr. Radioakt. Elektr.*, 2 (1905), 235.

[3] *Ibid.*, p. 265.

[4] O. Hahn, "A New Radio-active Element, which Evolves Thorium Emanation," *Nature*, 71 (1905), 574; cf. Lawrence Badash, *Rutherford and Boltwood* (New Haven: Yale University Press, 1969), pp. 57, 73.

[5] Badash, *Rutherford and Boltwood*, p. 56.

[6] *Ibid.*, p. 72.

[7] *Ibid.*, p. 81.

[8] *Ibid.*, p. 84.

[9] *Ibid.*, p. 88.

[10] *Ibid.*, p. 90.

[11] Thaddeus J. Trenn, *The Self-Splitting Atom* (London: Taylor & Francis, 1977), pp. 42–43.

[12] *Ibid.*, p. 44.

[13] *Ibid.*, p. 46.

[14] *Ibid.*, p. 49.

[15] *Ibid.*, App. 2.

[16] *Ibid.*, pp. 84–85.

[17] *Ibid.*, p. 115.

[18] Ernest Rutherford, *Radioactive Transformations* (New Haven: Yale University Press, 1906), p. 67.

[19] *Ibid.*, preface.

[20] *Ibid.*, p. 68; Friedrich Giesel had also independently found the substance named "radiothorium" by Hahn; cf. Hahn, "Über ein neues," p. 265, Soddy, "Radioactivity," p. 103.

[21] Rutherford, *Radioactive Transformations*, pp. 68–69.

[22] *Ibid.*, p. 69.

[23] *Ibid.*, pp. 68–69. This was also Hahn's opinion; see "Über ein neues," pp. 265–266.

[24] Rutherford, *Radioactive Transformations*, p. 68.

[25] Trenn, *Self-Splitting Atom*, p. 117.

[26] Rutherford, *Radioactive Transformations*, p. 69.

[27] Trenn, *Self-Splitting Atom*, p. 130.

[28] Rutherford, *Radioactive Transformations*, pp. 67, 172.

[29] Soddy, "Radioactivity," p. 164.

[30] Rutherford, *Radioactive Transformations*, p. 168.

ERNST H. BERNINGER

THE DISCOVERY OF URANIUM Z BY OTTO HAHN: THE FIRST EXAMPLE OF NUCLEAR ISOMERISM

Some discoveries in the history of physics were made too early – in any case, that could be the conclusion a historian of science might draw in hindsight. Whether or not that is the case with Otto Hahn's discovery of nuclear isomerism is a matter we shall return to later. In the meantime, let us think back to the period before 1921. Together with Lise Meitner, Hahn had identified a large number of radioactive substances and had determined their places in the radioactive series. With the discovery of protactinium in March 1918, the last gap was filled. Hahn and Meitner then published the following summary of their findings in the periodical *Physikalische Zeitschrift*:

1. The up-to-now hypothetical parent substance of actinium has been discovered, and produced in a radioactively pure state, in a concentrated form in rare-earth acids. It is a higher-order homologue of tantalum.
2. It emits alpha rays with a penetration range of 3.14 cm.
3. Its half-life is between a minimum of 1200 years and a maximum of 180,000 years.
4. The ensuing formation of actinium has been proven:
   (a) by plotting alpha-ray curves,
   (b) by measuring the actinium emanation, the increase in which we have been able to follow day by day for months now, and
   (c) by the radioactive precipitation, wich we were able to collect in increasing amounts on negatively charged plates.
5. In addition, the Curie value for the half-life of actinium was confirmed.
6. The new radioactive element has been named "protactinium."[1]

Although the work in this field was now completed, the position of protactinium in relationship to uranium was still unknown. Hahn conjectured there was a branching-off, with the newly discovered uranium Y as a decay product of uranium and as the parent substance of protactinium.

On January 21, 1921, he wrote in *Die Naturwissenschaften*:

Since the discovery of protactinium, the parent substance of actinium, there are no longer any empty slots in the three major series of radioactive elements, and it may be considered out of the question that any active material might still be found which could fit in any direct succession in one of the above-mentioned series. It must be recalled that no independent existence is attributed to the actinium series, which is rather considered to be a so-called sideline of the uranium-radium series, while the thorium series stands independent of uranium.[2]

William R. Shea (ed.), Otto Hahn and the Rise of Nuclear Physics, 213–220.
Copyright © 1983 *by D. Reidel Publishing Company.*

How could he find a confirmation of his surmise? In this situation Hahn struck out on a path that was typical of him. First he checked over his papers on the work he had done on uranium Y in 1914.[3] He came upon the fact that the results of the electroscopic measurements, which he had carried out with Lise Meitner, had revealed certain small inconsistencies. At the time, Hahn had not attached any particular importance to those findings. However their significance now increased in the light of his new concern with establishing a connection between uranium Y and protactinium. Furthermore, there was a hypothesis speculating that a new isotope of uranium might have to be considered the starting point of the actinium decay series. It would also be possible to clear up another inconsistency at the same time: the chemically determined atomic weight of uranium was then 238.2; however, if one calculates from the atomic weight of 226 for radium, formed after three alpha-decay processes, one arrives at the atomic weight of 238 for uranium.

The assumption of an actino-uranium (240) would have served to clarify matters. Special importance was placed on this hypothesis since Rutherford's discovery of a helium isotope with the mass of 3. Those were enough reasons for Hahn to start searching for a new radioactive substance after all, although − as he had stated − all the slots in the various decay series were already filled. The material he used as a starting point for his separations was uranium. He first turned his attention to uranium Y, which he separated from its parent substance, uranium X, quantitatively by means of repeated iron precipitation, according to the method published in 1914.[4] However, with the amounts investigated, it was not yet possible to determine with certainty whether an unknown substance was responsible for deviations in the half-life figures, that is, in the electroscopic measurements.

Matters were different with protactinium. In this case, Hahn mixed uranium salts with a hydrofluoric solution of small amounts of tantalic acid. The tantalum was then precipitated out. This procedure led to the precipitation of the protactinium contained in the uranium. When this work was carried out rapidly enough, Hahn observed besides the uranium X another weak, diminishing activity. The origin of this radiation was unknown to him at first. By subtracting the radiation of the uranium X from the measured radiation, he was able to set the half-life at between 6 and 7 hours. With a half-life of this duration for the unknown substance, it was important to exclude the possibility that the measurements were registering an effect caused by a minute contamination with the actinium isotope, mesothorium II. The half-life of mesothorium II was already known to be 6.2 hours.

Such a suspicion was quite a real possibility for Hahn, for at that time his institute, the Kaiser Wilhelm Institute for Chemistry in Berlin, was working with sizable amounts of mesothorium. Yet the chemical properties of the unknown substance spoke against its being mesothorium. Together with Lise Meitner, Hahn had demonstrated in 1919 that when actinium is treated with fluoric acid the actinium remains in the insoluble residue in the fluoric acid.[5] Mesothorium II and actinium are isotopes; it was therefore to be expected that mesotherium II would act the same way chemically. Hahn also took steps to exclude any contamination of the laboratory receptacles. Besides new glassware, brand new platinium crucibles were used.

After Hahn had precluded any possibility of contamination in this way, by repeating the experiments he was able to prove that the substance in question was a new radioactive product. He named this new "element" uranium Z. This uranium Z was undoubtedly an isotope of protactinium. Hahn proceeded to determine its radioactive properties, first producing pure preparations of uranium Z from large quantities of uranium. The result of measurements in a beta-ray electroscope was a half-life of 6.7 hours, allowing for a 2 percent margin of error. Hahn then also attempted to determine the proportion of uranium-Z radiation in the entire radiation. He arrived at a value of around 0.25 percent. He himself considered this only to be an approximation, for there was the difficulty of determining the proportion of radiation of uranium Z to the complex uranium X, which itself consists of two beta-radiating components, UX_1 and UX_2.

After the absorption coefficient of uranium Z in aluminum had also been determined, the new radioactive "element" was considered "identified," according to the standard criteria at that time. However, an important question still remained: what was the parent substance of uranium Z? Every attempt at finding an answer proved extremely difficult. There were two possibilities, wrote Hahn on March 14, 1921, in a published report:

Only UX_1 or a new UX_1 isotope with a similar life period comes into question as the parent substance. In the first case, UX_1 would undergo a dual decay in a manner which up till now has never been observed in a radioelement. In the second case, the most plausible assumption would be the existence of a new uranium series of a lower radiation intensity, whose individual members could be classified as isotopes in the known uranium-radium series.[6]

The very order in which Hahn presented these two possibilities doubtlessly indicates that he himself held the first solution to be more likely. Yet this was precisely the one which had to be considered out of the question in those days, for UX_1 and U II had been established as itegral, well-defined entities.

One would therefore assume that the intermediate product as well would be a single substance.

In his memoirs Hahn wrote: "Following the publication of the results of my extensive and nowise easy investigation, I took a ski trip with my wife in early 1921 to St. Christoph am Arlberg in Austria. . . . My work in the institute was interrupted a second time in 1921 by my having to be in the hospital from the 1st of August to the 1st of October."[7] But Hahn persevered in his search for the parent substance of uranium Z. First he ascertained the proportion of uranium X to uranium Z in uranium X of various ages, which he had extracted from several kilograms of uranyl nitrate. Within a certain margin of error, the proportion of UX to UZ remained practically constant. In other words, uranium Z seemed to decrease in direct proportion to uranium X. This fact seemed to speak in favor of uranium X_1 possibly being the parent substance of uranium Z. However, it was not possible to carry on the experiment long enough for a definite determination. The intensities weakened so fast, along with the diminishing uranium X, that the tests could be pursued for only about two months.

Hahn finally resolved to undertake a laborious large-scale experiment, which he began with 100 kilograms of uranyl nitrate.[8] The separation of a uranium-X preparation required several weeks. Finally he had highly active uranium X at his disposal, in a strength equivalent to approximately 50 kilograms of uranyl nitrate. Using this preparation, he again carried out uranium-Z tests. The final test was made when the uranium X was 156 days old, that is, having undergone disintegration down to 1.07 percent of the original amount. The results are given in the table shown in Figure 27, which records the yield of uranium Z from uranium X at various ages. The first column gives the age of the uranium X in days since it was separated from uranium. The second column shows the percent of uranium X still present at that time. The third column shows the yields of uranium Z in the beta-ray electroscope fitted with a 0.03 millimeter-thick aluminum foil. The figures given show the proportion, per thousands, of uranium Z to uranium X. And finally, the fourth column presents the yields in the electroscope with a 0.07-millimeter-thick aluminum foil.

In the case of the thicker foil, the yield is less because in this electroscope the uranium Z gives much poorer readings. The results can be seen in the graph in Figure 27. Above is the activity through a 0.03-millimeter-thick aluminum foil; below, the activity through a 0.07-millimeter-thick aluminum foil. The broken lines show the theoretical courses of uranium Z yields if its parent substance were not uranium X, which decreases in 24 days, but some

Ausbeuten an Uran Z aus verschieden altem Uran X.

Alter des UX in Tagen	Menge des zu dieser Zeit noch vorhandenen UX	Aktivität durch 0.3 mm Al UZ: UX in °/oo	Aktivität durch 0.07 mm Al in °/oo
2	94.0	2.67	1.14
4	88.8	2.68	1.17
5	86.5	2.44	1.07
12	70.6	2.70	1.19
15	64.8	2.80	1.14
18	59.4	2.70	1.15
53	21.7	2.85	1.27
82	9.4	2.38	1.14
97	6.0	3.02	1.34
136	1.95	2.79	1.18
156	1.07	(1.13)	–
		Mittel 2.70	Mittel 1.18

Uran Z-Ausbeuten aus verschieden altem Uran X.

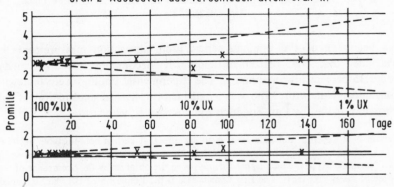

Fig. 27. Yield of uranium Z from uranium X at various ages. (From Otto Hahn, "Über das Uran Z und seine Muttersubstanz," Z. Phys. Chem., 103 (1922), 471, Table 1 and Fig. 1.)

unknown isotope of uranium X with a half-life which differs from that of uranium X by ± 3 days. The upper broken lines indicate the yield for a hypothetical parent substance of uranium Z with a half-life of 27 days; the lower line, for one with a half-life of 21 days.

Otto Hahn came to the conclusion that a new isotope of uranium X as the parent substance of uranium Z would have to have a half-life that only deviates from that of known uranium X by 2 to at the most 3 days. Experience showed this to be highly unlikely. At the time it was not possible for him to achieve an even narrower delimitation by the utilization of even stronger preparations. Therefore, on October 21, 1922, Hahn sketched out the diagram shown in Figure 28 for the radioactive disintegration at

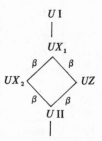

Fig. 28. Pattern of radioactive disintegration at the beginning of the uranium series. (From Otto Hahn, "Über das Uran Z und seine Muttersubstanz," *Z. Phys. Chem.*, **103** (1922), 472.)

the beginning of the uranium series. From absorption measurements he concluded that 99.65 percent of uranium X_1 decomposes into uranium X_2, and 0.35 percent into uranium Z. The experimental findings were therewith concluded. Not until 1936 was a satisfactory physical explanation of this diagram provided by Carl Friedrich von Weizsäcker, who wrote:

Certain new experimental data lead us to surmise that certain beta-unstable nuclei of a given charge and mass are not completely identical with each other, but appear in two "isomeric" types, which signal themselves to the observer by the occurrence of two different decay constants. The first known case is the isomerism of UX_2 and UZ, made exceedingly plausible by Hahn . . . . In order to understand the occurrence of UZ in the natural uranium series, one must assume that after the beta decay of UX_1, at least in those 0.35% of the disintegration processes in which UZ is produced, the rest of the nucleus remains in a stimulated state, from which the postulated "cascade leap" is possible.[9]

Let me return now to the question posed at the beginning. Starting with his work with Rutherford in Montreal, Hahn step by step gained the experimental proficiency which finally enabled him to work with micro-concentrations. The discovery of uranium Z marked a certain high point in Hahn's career. He himself considered it to have been his finest achievement, and on occasion he jestingly termed it "a bit worthy of a Nobel Prize." In these projects, Hahn had to be able to detect amounts of radioactive substances as small as 10^{-14} gram. The high degree of confidence with which he could rely on his measurements and his keen ability to classify the results show the measure of his qualifications. In the years following the discovery of uranium Z, Hahn created the professional field of applied radiochemistry.

At the beginning of the 1930s when Enrico Fermi, in his experiments with neutron irradiation of heavy elements, came to an element with the atomic number 93, a co-worker of Hahn's, Aristid von Grosse, conjectured that this element 93 was in reality protactinium. Hahn's interest in these questions was now rekindled: after all, he and Lise Meitner were the greatest experts on protactinium, which he had discovered. Hahn and Meitner repeated Fermi's experiments, and Hahn later wrote about the results:

We carried out a so-called indicator experiment with with a β-radiating isotope of protactinium, the uranium Z discovered by me some 12 years earlier, so as to be completely sure. . . . The outcome of the indicator experiment was completely faultless. The Fermi elements were quantitatively separated from element 91. There is no alternative to classifying the new Fermi substances as element 93 or as 93 and 94. And these substances were precipitable from a strongly hydrochloric solution with H_2S, exactly analogous to the lower homologue of rhenium and the platinum metals following it.[10]

In the years 1935 to 1938, Hahn, Meitner, and Fritz Strassmann put forward the transuranic series for consideration and explanation. If we consider that von Weizsäcker's interpretation of nuclear isomerism brought with it a final confirmation of the uranium Z studies, it is quite understandable that when the three isomeric series arising from uranium 92 were still mystifying the physicists, Otto Hahn – the chemist – simply accepted them. After all, his experimental work dating back fifteen years did indeed receive a brilliant theoretical interpretation. Taking that into consideration, it becomes understandable how Hahn and Meitner could have held so firmly and so long to their conviction regarding the "false transuranic series."

In conclusion, I would like to call attention to another question: in recent times, doubt has occasionally been cast on whether Hahn deserves the main credit for splitting the atom. Sometimes the assertions even claim that it was mainly Fritz Strassmann's experimental proficiency which led to the success. Strassmann's role in the discoveyr of nuclear fission is undisputed; Lise Meitner's preparatory contributions were also an important step on the road to the discovery. But I would like to emphasize that the exemplary manner in which Otto Hahn had already basically applied the art of experimentation and the methodical interpretation of the results during the discovery of uranium Z in like measure later led to the discovery of nuclear fission.

Deutsches Museum

NOTES

[1] Otto Hahn and Lise Meitner, "Die Muttersubstanz des Actiniums, ein neues radioaktives Element von langer Lebensdauer," *Phys. Z.*, 19 (1918), 208–218.

[2] Otto Hahn, "Über ein neues radioaktives Zerfallsprodukt im Uran," *Naturwissenschaften*, 9 (1921), 84.

[3] Otto Hahn and Lise Meitner, "Über das Uran Y," *Phys. Z.*, 15 (1914), 236–240.

[4] Otto Hahn and Lise Meitner, "Nachweis der Existenz von Uran Y," *Phys. Z.*, 15 (1914), 236–237.

[5] Otto Hahn and Lise Meinter, "Über das Protaktinium und die Frage nach der Möglichkeit seiner Herstellung als chemisches Element," *Naturwissenschaften*, 7 (1919), 611–612.

[6] *Ber. Deut. Chem. Ges.*, Abt B, 54 (1921), 1142.

[7] Otto Hahn, *Mein Leben* (Munich: Bruckmann, 1968), p. 135.

[8] Otto Hahn, "Über das Uran Z und seine Muttersubstanz," *Z. Phys. Chem.*, 103 (1922), 461–480.

[9] Carl Friedrich von Weizsäcker, "Metastabile Zustände der Atomkerne," *Naturwissenschaften*, 24 (1936), 813–814.

[10] Otto Hahn, "Die 'falschen' Transurane. Zur Geschichte eines wissenschaftlichen Irrtums," *Naturw. Rund.*, 15 (1962), 44.

NUCLEAR PHYSICS IN CANADA IN THE 1930s

Nuclear physics in Canada in the 1930s meant principally radioactivity of the naturally occurring heavy elements. It is appropriate to note that Otto Hahn, employing radiochemical methods, discovered a number of these radioactive substances: in particular, radiothorium (^{228}Th), radioactinium (^{227}Th), mesothorium 1 (^{228}Ra), mesothorium 2 (^{228}Ac), protactinium (^{231}Pa), uranium Z (^{234}Pa), and several in the active deposits. He also contributed greatly to our detailed knowledge of the alpha, beta, and gamma radiations from nuclei. While this paper deals mostly with such radiations, neutrons must also be mentioned. The discovery of radioactive fragments from the neutron-induced fission of uranium by Hahn and Strassmann at the end of the 1930s triggered off a new variety of nuclear physics, but that is beyond the time scope of interest here.

The constraints imposed by the title of this chapter will now be emphasized. Certainly, research work of international standard in fields other than nuclear physics was done at several universities in Canada in the 1930s. For example, at the University of Toronto the important fields of research were atomic spectra and atomic structure, molecular spectra, and low-temperature physics, especially superconductivity and the properties of liquid helium. The use of hyperfine structure of spectral lines to determine mechanical and magnetic moments of nuclei is not regarded here as a part of nuclear physics, since the experimental technique involved optical interferometers rather than radiation detectors. As another example of non-nuclear research, McGill University enjoyed a monopoly on the Stark effect. My selection of nuclear research will be confined to four universities: Queen's, Dalhousie, Saskatchewan, and Laval.

As we go forward, you will often note the Rutherford connection. Professors who studied as research students under Rutherford at Manchester or Cambridge in turn directed research students in Canada, who went on to study under Rutherford at Cambridge. There were in Canada what we might call "two generations" with the Rutherford connection. I am one of the surviving members of the "second generation." But there were exceptions to the Rutherford connection; these will also be noted in what follows.

Accomplishments in the 1930s must be weighed against the resources that

William R. Shea (ed.), Otto Hahn and the Rise of Nuclear Physics, 221–240.

were available — financial, radioactive, and instrumental. The Great Economic Depression, which began with the crash of the stock market in October 1929, continued until deep into the 1930s. The plight of the National Research Council of Canada in the period 1930–1935 is indicative of the lack of financial resources in support of scientific research. The consequences of the prime cause were intensified by a hostile Minister and a so-called sound money policy of the Conservative Government in power. The program of NRC scholarships for graduate students, which was nicely underway in the 1920s, had to be cut back to one-fifth in money available annually. The NRC grants for assisted researches in the universities had also to be slashed. In desperation, NRC was driven to allowing student scientists to work *without pay* in its research laboratories.[1]

Under the circumstances the Canadian universities had to provide funds, however meagre, for research on campus. For example, at Queen's University some early preparation had been made. A Committee on Scientific Research had been formed in the winter of 1916–1917 by action of the Board of Trustees; it was intended to play a role in the university similar to that played in Canada as a whole by the National Research Council, which had been formed in November 1916. The committee administered a fund for equipment and supplies and for the summer salaries of research students. It was also possible to hire bright undergraduates in research projects during the summers. However, a grant of $1000 was regarded as large. Since junior academic staff had no chance whatever of obtaining a research grant from NRC in the 1930s, this local support was invaluable.

For studies of ranges and energy spectra of beta particles, sources had to be obtained from natural materials. The most exciting results turned out to be the special properties of the disintegration electrons. As we shall see, radium E (^{210}Bi) and uranium X_2 (^{234}Pa) proved to be almost ideal sources. The long suspense over the energy spectrum of the true disintegration electrons is perhaps well known. Here, it is amusing to note that Rutherford made a speculation that turned out to be incorrect: In his 1913 book, after reviewing the experimental evidence which showed that all beta emitters investigated, with two exceptions, yielded sharp lines in magnetic spectrographs, he wrote: "In the light of the results given above it appears not improbable that the continuous beta ray spectrum observed for uranium X and radium E may be ultimately resolved into a number of lines."[2] One year later, James Chadwick showed that the continuous energy spectrum is the rule, not the exception; it is indeed the spectrum of the true disintegration electrons.[3] The sharp lines photographed in magnetic spectrographs were caused by (what we now call) "internal conversion electrons." They arise from nuclear de-excitation following the emission of the true disintegration electrons and are associated with the emission of gamma rays.

Nature was very kind in providing two nearly ideal sources of very fast disintegration electrons. The energy spectrum of radium E extends up to 1.16 million electron volts (MeV), that of uranium X_2 to 2.3 MeV. These disintegration electrons have therefore relativistic velocities extending up to 98 percent of the velocity of light. The precursors, radium D (^{210}Pb) and uranium X_1 (^{234}Th), have conveniently long half-lives, 22 years and 24 days respectively, and emit only very low energy beta and gamma rays. In general, it is therefore advantageous to use radium E and uranium X_2 in equilibrium with their precursors.

In the 1930s a number of nuclear physicists collected old radon tubes from hospitals. These tubes had been filled with radon (^{222}Rn) for therapeutic treatments with gamma rays in radiology departments. Several hospitals in Canada had radium (^{226}Ra) in solution, from which from time to time radon was compressed into glass capillary tubes and sealed off. The long-lived radium (D + E + F) could be readily extracted from the ground-up tubes, and, if necessary, separated one from another by electrochemical methods. The most productive use of such source material in the history of physics was in experiments done by James Chadwick in his discovery of the neutron. The very strong radium F (^{210}Po) source of alpha particles which he used was extracted from old radon tubes from a Baltimore hospital. The personal connections had been made by Norman Feather when he spent the year 1929–1930 at Johns Hopkins University.[4]

The separation of uranium X_1 from a uranium compound was more troublesome and time-consuming. At Queen's University one started with several kilograms of "yellow cake." The ultimate source had to contain essentially the whole uranium-X activity in the "yellow cake," and to be as free as possible of extraneous solid matter — say, down to a few milligrams. In general, graduate students in Professor J. A. Gray's laboratory had to make their own sources. Otto Hahn would have been amused if he had seen physics students doing radiochemistry.

It is a matter of some historical interest to note that the principal radiation from uranium which blackened Henri Becquerel's photographic plates in his discovery of radioactivity was the energetic beta particles of uranium X_2. The excellent radiograph of an aluminum medal which he took in March 1896 is proof enough of my statement.[5] The work on the beta particles to be described was done partly at Queen's University under the direction of

Professor Gray and partly at the Cavendish Laboratory by his former students. There was a certain amount of coming and going by research students between the two laboratories.[6] A closely integrated story will be attempted.

J. A. Gray was an Australian who took first-class final honors in mathematics and in physics in consecutive years at the University of Melbourne. In 1905 he won a book prize, which allowed him to select four books. One of these was Ernest Rutherford's *Radio-activity*, which had been written the year before at McGill University. This book determined the main course of Gray's life.[7]

When Gray failed to be chosen in 1908 for an Exhibition of 1851 Science Research Scholarship to the University of Cambridge, he went to England anyway and worked for one year on radioactivity under the Honourable R. J. Strutt at the Imperial College of Science. On being renominated in the following year by the University of Melbourne, with three publications to his credit, Gray was awarded an Exhibition of 1851 Scholarship, which he held for three years in Rutherford's laboratory at the University of Manchester. In 1912 Gray came to Canada and joined the Physics Department of McGill University. After serious interruption of his research work by enlisting and serving in World War I, and after further years at McGill, he was appointed to the Chown Science Research Professorship at Queen's University. He arrived at Queen's in January 1924.

The Exhibition of 1851 Science Research Scholarships were prestigious and were good for travel overseas and for two or three years of study and research; in these respects they were unique until after World War II. Three were awarded annually to Canadian students in the sciences. Dalhousie, McGill, and Queen's University made a practice of nominating their outstanding graduate students for these scholarships. There was an obvious strategy in finding and preparing candidates. The principal requirement was a thesis of exceptional quality after two or three years of research work. By this means of support a number of Gray's students were able to work under Rutherford at the Cavendish Laboratory.

The decade of the 1930s was a time for the rapid development of more sensitive instruments for detecting beta and gamma rays. At Queen's University Gray had a continuing program for upgrading his detection instruments. When he moved to Kingston in January 1924 he brought the following tools of his profession: several stopwatches and an assortment of gold-leaf electroscopes in the form of cubical boxes made of brass, lead, tin, and aluminum. When I returned to Queen's in 1930 after two years at the Cavendish Laboratory, I graduated from gold-leaf electroscopes to a Compton electrometer. The Geiger-Müller discharge tube with scale-of-two counter was introduced by W. J. Henderson upon his return from Cambridge. However, such discharge tubes were not yet fully reliable, and the main thrust of Gray's program was toward new types of ionization amplifiers with sensitive galvanometers. J. S. Marshall contributed greatly to the facilities by developing and building two ionization amplifiers based on the General Electric Company's Pliotron FP-54 vacuum tube. This was a four-element tube especially designed for the measurement of small direct currents. The positive-ion collecting electrode in the ionization chamber was directly connected to the control grid of the Pliotron. When this insulated system was disconnected from ground, the ionization current was measured by the rate of deflection of a sensitive galvanometer in a circuit designed to maintain a stable balance under uncontrolled conditions.[8] In 1940 H. Le Caine and J. H. Waghorne invented and built a new type of ionization amplifier.[9] Whenever electric charge was collected by the central electrode of the ionization chamber, an alternating voltage was obtained by electrostatic translation of the motion of a reed. This was amplified by an a.c. amplifier and used to measure the ionization current by a balance method involving automatic timing.

With the development of the sensitive detectors by the graduate students named above, and with the production of strong radioactive sources through radiochemistry by graduate students, especially J. F. Hinds and A. G. Ward, it became possible to experiment with *monoenergetic* beta particles selected by semicircular focusing in the uniform field of an electromagnet. A very careful set of experiments on the transmission of essentially monoenergetic beta particles through aluminum was done by J. S. Marshall and A. G. Ward.[10] Although these were not the first of their kind, the older sets disagreed greatly with each other. Figure 29 shows a sample of the higher-energy results. The ionization of the transmitted beam is usually plotted against the mass m (in gm/cm^2) of aluminum through which the beta particles have passed. Here, in fulfilment of the old expectation that the shapes of the curves are the same, the ionization is plotted against m/R, where R is the extrapolated range determined by the intercept of the linear section of the curve with the axis of mass per unit area. The sets of points for the three energies are in very close agreement.

In 1939 I proposed a useful formula for the extrapolated range, or simply the range, R, as a function of the energy E, for beta particles of energy greater than 0.6 MeV:[11]

$$R = 0.526E - 0.094.$$

226   B. W. SARGENT

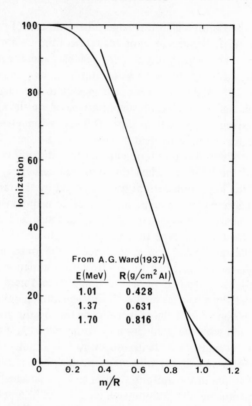

Fig. 29. Universal transmission curve for monoenergetic beta particles through aluminum.

The transmission curves by Marshall and Ward and this formula may be found in several textbooks. Figure 30, from my work, shows examples of the transmission curves observed when the whole beta spectrum is incident on the aluminum absorbers. The ionization caused by the beta and gamma rays falls rapidly while the beta particles are being slowed down and stopped; after the mass per unit area exceeds the range R_{max} of the fastest beta particles, the ionization by the gamma rays alone falls less rapidly.

The actinium (^{227}Ac) preparation which I used had an interesting "ancestry." Upon request, I received it as a gift in September 1935 from Professor Aristid von Grosse of the Chemistry Department of the University of Chicago. At one time he had been a colleague of Otto Hahn. The actinium preparation

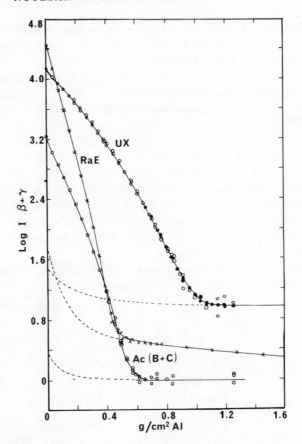

Fig. 30. Transmission curves for beta and gamma rays through aluminum. On notation: AcC includes its descendant AcC'' and UX is short for UX_1 and UX_2 in equilibrium. (From B. W. Sargent, *Can. J. Res.*, A17 (1939), 82. Reprinted with the permission of the National Research Council of Canada.)

had been removed from radioactive material being purified by von Grosse as a preliminary to his determination of the atomic weight of protactinium.[12]

When I went to the Cavendish Laboratory in 1928 my immediate supervisor, C. D. Ellis, assigned me the general problem of the energy spectra of the disintegration electrons, and the particular problem of the reality of the upper limits of energy or end-points. With the weak radioactive sources available, the only practical method of investigation was the absorption method. This I had learned very well under the supervision of Professor Gray. The range R_{max} of the fastest disintegration electrons in a spectrum was chosen by inspection from an absorption curve of the type shown in Figure 30. To continue with the illustrative examples of that figure, the values of R_{max} by inspection are approximately 0.5, 0.65, and 1.1 grams/centimeter^2 of aluminum for the beta particles of radium E (^{210}Bi), actinium (B + C'') (^{211}Pb + ^{207}Tl), and uranium X_2 (^{234}Pa) respectively. The available transmission curves for monoenergetic beta particles, which predated the results of Figure 29, were used for calibration, by which the end-point energy E_{max} corresponding to R_{max} was derived. In the collected results on the end-points, which were published in 1933, one conclusion was stated as follows: "Many experiments, designed to determine an end-point and to set an upper limit on the fraction of disintegration electrons which have greater energies, in particular cases, afford a strong indication that the end-point is one of the characteristics of the spectrum."[13]

When the end-point energy E_{max} was plotted against the decay constant λ_β, as in Figure 31, an empirical relation appeared to exist. The left-hand side

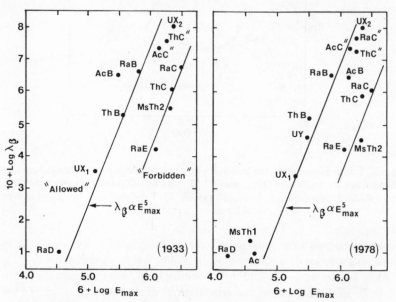

Fig. 31. Empirical relations between the end-point energy and the disintegration constant of beta emitters. (Adapted from B. W. Sargent, *Proc. Roy. Soc. Lond.*, A139 (1933), 659.)

of Figure 31 displays the original points of 1933; the right-hand side shows the points for the principal partial spectra taken from the latest edition of the *Table of Isotopes*.[14] The radioactive substances appear to fall into two groups, with the factor of at least 100 in λ_β between the lines, which have slopes indicating λ_β is proportional to E^5_{max}. The two groups have been named, with theoretical connotation, "allowed" and "forbidden" transitions.[15]

It was possible to extend the absorption method to yield the full energy spectrum of the disintegration electrons. By trial, an assumed shape of energy spectrum was combined with the existing transmission curves for monoenergetic beta particles until a calculated absorption curve for the whole spectrum was obtained in acceptable agreement with experiment. The collected results on the full energy distributions of the disintegration electrons were published in 1932.[16] The spectra for the disintegration electrons of radium B (^{214}Pb), radium C (^{214}Bi), radium E (^{210}Bi), and thorium B (^{212}Pb) were taken from earlier work with magnetic spectrometers, while those of uranium X_2 (^{234}Pa), thorium C (^{212}Bi), thorium C″ (^{208}Tl), and actinium C″ (^{207}Tl) were derived by the absorption method. These collected experimental results on energy spectra and end-points were published fortuitously in time to be used by Enrico Fermi in his theory of beta decay. His paper, published in the *Zeitschrift für Physik* in 1934, is now regarded as a classic.[17]

The validity of the conservation of energy principle for individual atoms, each emitting one disintegration electron, was the subject of some argument between Niels Bohr and others. The principle was quickly subjected to a numerical test by C. D. Ellis and N. F. Mott.[18] Thorium C (^{212}Bi) decays by alpha-particle emission to thorium C″ (^{208}Tl) and by beta-particle emission to thorium C′ (^{212}Po), which in turn decay to thorium D (^{208}Pb) by beta- and alpha-particle emission respectively. Using the end-point energies for the beta disintegration energies, Ellis and Mott obtained 11.20 and 11.15 MeV respectively by the two paths. Included in 11.20 MeV are gamma rays of energy 0.58 and 2.62 MeV from thorium C″. The numerical agreement is much better than if the average energies of the beta spectra had been used as the disintegration energies. Pursuing this matter further at the suggestion of Ellis, W. J. Henderson with great care determined the end-points of the beta spectra of thorium C and thorium C″ in a magnetic spectrometer.[19] The revised sums, 11.19 and 11.20 MeV by the respective paths, are in such good agreement that the conservation of energy principle is clearly valid for the end-point energies as the disintegration energies.

The paper by Ellis and Mott is important for more general reasons. They showed that beta-decay branching to excited states of the product nucleus can occur, with subsequent de-excitation by the emission of internal conversion electrons or gamma rays. The spectrum of disintegration electrons is then complex, being the sum of partial spectra with separate end-points. The conservation of energy principle is valid for the competing beta decays when the end-point energies are used as the disintegration energies to the excited states.

So far I have followed a phenomenological course and avoided mentioning Wolfgang Pauli's suggestion that a neutral particle, later named the neutrino by Fermi, is emitted simultaneously with the disintegration electron in beta decay.[20] Certain fundamental difficulties in understanding the process had arisen in the late 1920s and have been recorded by Pauli and others.[21] The microcalorimetric experiments by C. D. Ellis and W. A. Wooster and by L. Meitner and W. Orthmann gave 0.34 ± 0.02 MeV for the average energy emitted per disintegration of radium E.[22] From the direct counting of the beta particles per disintegration, in experiments which were inherently difficult, the number was not greater than one by more than the uncertainty of measurement.[23] Since the number of true disintegration electrons emitted per disintegration must be exactly one from the Fajans-Russell-Soddy displacement law in the periodic table of the elements, the number of secondary electrons per disintegration is negligible. The average heating per disintegration, 0.34 ± 0.02 MeV, is therefore the average energy of the true disintegration electrons from radium E. This value is the same as the average energy of the beta particles, determined experimentally in magnetic spectrometers, within the uncertainties of the two types of experiments. Secondary causes of the energy spread have thus been ruled out. The observed disintegration electrons appear to be deficient in kinetic energy by various amounts from zero up to 1.16 MeV. But this difficulty with a conservation principle is not the only one.

In beta decay the parent and product nuclei always have the same mass number, which is, of course, an even or odd integer. Both nuclei must have the same statistics (Fermi statistics if odd, Bose if even), and their nuclear angular momenta can differ only by zero or an integral multiple of $\hbar$ $(= h/2\pi)$. Since the disintegration electron, as a spinning electron, is a fermion with intrinsic spin $\frac{1}{2}\hbar$, an angular momentum of $\frac{1}{2}\hbar$ is missing in beta decay. The missing properties may be ascribed to a phantom. Pauli's postulate of the simultaneous emission of a neutrino with the disintegration electron in beta decay seemed inescapable if the conservation principles of energy, angular momentum, and statistics were to be saved. At that time there was no experimental evidence as to the validity of the conservation of linear momentum. The kinetic energy of the product nucleus recoiling in beta decay is so small (less than 100 eV) that the difficulties of measurement were formidable. However, some linear momentum would also have been missing. The sum of the kinetic energies of the disintegration electron and the neutrino is for all practical purposes always the end-point energy of 1.16 MeV in the decay of radium E.

Pauli included his neutrino postulate in an invited paper in a symposium on "The Present Status of the Problem of Nuclear Structure" at the Pasadena meeting of the American Physical Society in June 1931.[24] As he explained many years later,[25] the matter was so uncertain that he did not allow his talk to be printed. The first printed account of his ideas was in the proceedings of the seventh Solvay Congress held in Brussels in October 1933.[26] The first printed account which became available to me was in Fermi's theoretical paper of 1934.[27] My papers written in 1932 were done without the knowledge of Pauli's hypothesis of neutrino emission; this is a matter which I still regret.

Fermi's theory of beta decay was based on the neutrino postulate. Nuclear constituents were assumed to be protons and neutrons only. The electron and neutrino were created at the moment of emission when a neutron transformed to a proton. The shape of the energy spectrum of the electrons and the relation of the decay constant with the end-point energy were explicitly derived for an "allowed" transition. The empirical results in Figure 31 were thereby given theoretical explanations. The division between "allowed" and "forbidden" transitions was seen to follow from certain selection rules. For high values of E_{max} in "allowed" transitions, the theory shows that λ_β becomes proportional to E_{max}^5.

At the Cavendish Laboratory, W. J. Henderson compared his experimental spectra of thorium C and thorium C″ near the end-points with Fermi's formula and concluded that the mass of the neutrino must be much less than the rest mass of the electron and possibly zero.[28] This confirmed Fermi's tentative conclusion from the high-energy tails of the old energy distributions.

Progress on the theoretical explanation of the beta spectrum and on the interrelation of partial spectra and gamma rays increased the need for more and better measurements of end-points, especially of the new radioactive substances being made by nuclear reactions. The range method of determining E_{max} through R_{max} was recognized as simple and not requiring the

232   B. W. SARGENT

strength of source that would be needed in a magnetic spectrometer. The method of locating R_{max}, simply by inspection, as the point where the beta-ray ionization first comes to zero (as in Figure 30) was recognized as susceptible to hidden errors. Two factors oppose each other near the end of the range. Figure 29 shows that beta-ray ionization is detectable well beyond the extrapolated range. Consequently, if a very strong source is used, R_{max} by inspection in Figure 30 will move to the right. Secondly, a kind of "signal-to-noise ratio" enters; if it is poor owing to low strength of source or high gamma-ray background, R_{max} by inspection will move to the left. In 1938 Norman Feather and I independently proposed a comparison method which allows an end-point to be established to 2 percent.[29] My work will be described.

The beta-ray spectra of radium E and uranium X_2 were chosen as standards. Their end-points had been carefully determined by semicircular focusing in a magnetic spectrometer by A. G. Ward and J. A. Gray.[30] Energies of 1.15 and 2.32 MeV and ranges R_{max} of 0.51 and 1.12 gm/cm^2 of aluminum respectively were adopted. The measured ionization currents of the beta particles only, when plotted on a logarithmic scale against m/R_{max}, have the same shape for both beta emitters above $m/R_{max} = 0.5$, and can be made coincident as in Figure 32. The best value of R_{max}, and hence E_{max}, of another spectrum can be found by trial to fit the common curve in Figure 32.

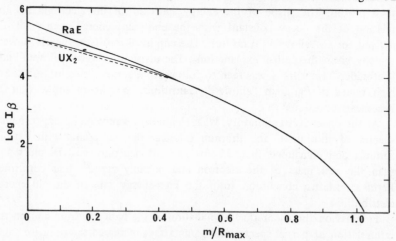

Fig. 32. Transmission curves for the beta rays of radium E and uranium X_2 through aluminum. (From B. W. Sargent, *Can. J. Res.*, **A17** (1939), 82. Reprinted with the permission of the National Research Council of Canada.)

The first applications were to the beta spectra of actinium B and actinium C'', which have half-lives of 36.1 and 4.79 minutes respectively. Actinium B was collected by radioactive recoil on aluminum foil exposed to actinon (^{219}Rn) in an air stream from the actinium preparation obtained from von Grosse. The method of radioactive recoil was exploited long ago by Hahn and Meitner for separating radioactive substances. In my work collection times down to 1 minute were used, and the ionization of the beta particles which had penetrated various thicknesses of aluminum was measured for 60 minutes. The composite absorption curve for actinium B and actinium C'' in equilibrium was shown in Figure 30. Using the growth curves of the activity, separate absorption curves were obtained for the beta particles of actinium B and actinium C''. The ranges R_{max} are 0.64 and 0.68 gm/cm^2 of aluminum, and the end-points E_{max} are 1.39 and 1.47 MeV respectively.

The reliability of the comparative absorption method for determining end-points of beta spectra is shown in Table I. Values deduced by this method in the late 1930s are compared in the last two columns with the best from magnetic spectrometers.[31] The most extensive use of my absorption method was undoubtedly that by B. L. Moore, working under the supervision

TABLE I

Comparison of the end-points of beta spectra measured by the absorption method with those measured much later by magnetic analysis and Kurie plots

Radioactive substance	Electron or positron	Half-life	E_{max} (MeV)	
			Absorption (1930s)	Magnetic analysis (1978)
AcB (^{211}Pb)	e$^-$	36.1 min.	1.39 ± 0.03	1.378 ± 0.008
AcC'' (^{207}Tℓ)	e$^-$	4.79 min.	1.47 ± 0.03	1.431 ± 0.008
^{11}C	e$^+$	20.7 min.	1.03 ± 0.03	0.961 ± 0.003
^{13}N	e$^+$	9.95 min.	1.22 ± 0.03	1.190 ± 0.003
^{24}Na	e$^-$	14.9 hr.	1.36 ± 0.05	1.389 ± 0.001
^{27}Mg	e$^-$	10.0 min.	1.74 ± 0.05	1.754 ± 0.004
^{32}P	e$^-$	14.4 days	1.72 ± 0.03	1.711 ± 0.002
^{76}As	e$^-$	26.3 hr.	3.12 ± 0.10	2.970 ± 0.002
^{104}Rh	e$^-$	41.8 sec.	2.46 ± 0.10	2.44
^{116}In	e$^-$	54.1 min.	0.95 ± 0.05	1.00
^{128}I	e$^-$	25.1 min.	2.08 ± 0.10	2.120 ± 0.010
^{198}Au	e$^-$	2.76 days	0.94 ± 0.05	0.961 ± 0.001

of R. F. Bacher at Cornell University.[32] Moore's results are for ten electron and positron emitters produced artificially. They have been included below the actinium B and actinium C″ end-points in Table I. The agreement with present-day best values is very good.

At Dalhousie University Professor G. H. Henderson and his graduate students established a distinguished record in radioactivity by alpha-particle emission. As an undergraduate he had studied in Professor H. L. Bronson's department, taking the double degree of Bachelor of Arts and Science with high honors in mathematics and physics and winning the highest award – the Governor General's Gold Medal. Two years later, in 1916, as a graduate student, he received the M.A. degree and won an Exhibition of 1851 Science Research Scholarship. After the war ended, his deferred scholarship enabled him to realize his long-cherished ambition of going to Cambridge to work with Rutherford.[33]

At the Cavendish Laboratory Henderson's measurements of the ranges of alpha particles in air, with improved precision over earlier work, raised the question of the electric charge on the alpha particle when it has slowed down and almost reached the end of its range in matter. This charge, which is normally two positive units, might fluctuate between two, one, and zero by the capture and loss of electrons from the atoms traversed. To say that Henderson was a pioneer in his experimental investigations of the phenomenon is not too strong a statement. After leaving the Cavendish Laboratory, he continued with this work during 1922–1924 while he was an assistant professor at the University of Saskatchewan. The interest in radioactivity that he aroused there was not wholly lost when he resigned and returned to the Physics Department of Dalhousie University. The capture and loss of electrons by heavy positive ions travelling through matter is now a lively subject, both for scientific and technological reasons.

In the late 1920s Henderson and his graduate students G. C. Laurence and J. L. Nickerson determined the ranges of the alpha particles from the very long-lived radioactive substances: uranium I (^{238}U), uranium II (^{234}U), and thorium (^{232}Th).[34] The technical problems arising from the shortness of the ranges and the low specific activity of the source material were minimized by using a large area and photographing the individual tracks in air in a small cloud chamber. The Geiger-Nuttall relation between the half-lives and the ranges of the alpha particles in the three radioactive series was extended more authoritatively to the very long half-lives, which enter into estimates of geological ages. Laurence's investigation won for him an Exhibition of 1851 Scholarship by which he was able to follow in Henderson's footsteps to the

Cavendish Laboratory to work on the alpha particles, which were Rutherford's favorite radiation.

In the latter half of the 1930s Henderson and graduate students S. Bateson, L. G. Turnbull, C. M. Mushkat, D. P. Crawford, and F. W. Sparks performed a classic series of experiments, using a halo photometer for the direct recording of dark ring patterns called pleochroic halos in mica.[35] In three dimensions, these halos are sets of tiny concentric spheres seen under a microscope. Since a beam of monoenergetic alpha particles makes a Bragg ionization curve with a pronounced maximum in the ionization just before the end of the range, the colored sphere is attributed to maximum radiation damage caused by the alpha particles from a radioactive inclusion at the center of the sphere. A very thin cleaved section passing diametrically through the spheres shows a set of concentric rings, one for each group of alpha particles. Figure 33 on the left-hand side shows photomicrographs of halos from inclusions identified as uranium and thorium.[36] The radii extend from 12 to 42 micrometers in mica. On the right-hand side of Figure 33 are

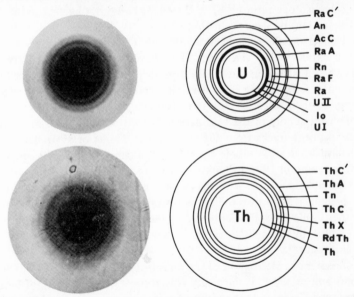

Fig. 33. Photomicrographs of pleochroic halos of uranium and thorium. (From G. H. Henderson and F. W. Sparks. *Proc. Roy. Soc. Lond.,* **A173** (1939), 238. Reprinted with the permission of the Royal Society and of Mrs. G. H. Henderson, Professor Nancy E. Henderson, and Mrs. F. W. (Henderson) Crickard.) For better definition of the rings, see the plate in the *Proceedings.* The schematic circles have been added.

schematic circles, drawn to the same scale as the photomicrographs, which identify the alpha-particle emitters through the ranges measured with the photometer. The rings, which have been forming over a time of 10^9 years, are quite sharp. The radii when converted to air equivalent agree well with the present-day ranges of the alpha particles. These two facts are usually taken as proof to a reasonable degree that the phenomenon of radioactivity has not changed in geological time.

Henderson and Bateson proposed a theory of halo formation, which was based on the summed Bragg curves for alpha particles and on artificial blackening experiments where the mica was exposed to known numbers of alpha particles from radon and its products.[37] Their calculated curves for blackening agreed fairly well with many of the photometer records. This theory was applied by Henderson to the ratio of the blackening of the actinium C (^{211}Bi) and radium C' (^{214}Po) halos for the determination of the ages of the uranium inclusions in three specimens.[38] However, reliable values of the half-life of ^{235}U, which is the head of the actinium series, and of the present-day isotope ratio ^{238}U/^{235}U (= 139/1) did not become available for such calculations until a decade or more later.

In Henderson's large collection of mica (biotite) samples a number of so-called extinct halos were seen. These have only one, two, or three rings, which were identified as due to alpha particles from certain later members of the uranium series having short lives on a geological time scale. He advanced a theory of their origin.[39] With the advantage of hindsight, it is easy to say that the central core of the halo pattern for the uranium series must contain fission fragment tracks from the spontaneous fission of ^{238}U. These tracks could have been made visible by chemical etching. However, there was no scientific reason in 1934 to look for such tracks, which are quite unlike the tracks of alpha particles. With the recent discoveries of giant and other new kinds of radioactive halos, the subject is once again alive.[40] The use of mica and other insulating solids for recording the tracks of heavy ions, individually made visible by chemical etching, is now a well-established technique in nuclear physics.[41]

Professor E. L. Harrington of the University of Saskatchewan was born in Missouri and received his university education at the University of Missouri, Harvard, and Chicago. At the University of Chicago he measured the viscosity of air as part of R. A. Millikan's program for the precise determination of the charge on the electron in the celebrated oil-drop experiments. Being very much interested in medical physics, Harrington was asked by the Saskatchewan Cancer Commission to build a radon plant. Skilled as a glass-blower, he built a good plant at the university and directed its operation until his retirement. He had access to the radon for his own nuclear-physical investigations to be mentioned below.[42]

With collaborators O. A. Gratias and E. O. Braaten, Harrington investigated the formation and mobilities of radioactive aggregates in gases containing polar molecules and radon.[43] One experimental method employed an ultra-microscope. The active deposit atoms tended to collect in aggregates on which water molecules, being polar, readily formed adsorbed layers. The behavior of these aggregates in an electric field was reminiscent of Millikan's oil drops in capturing electrons. The mobility of the molecular aggregates did not have a definitive value. At the end of the 1930s Harrington and J. L. Stewart measured the capture cross-sections for thermal neutrons of lithium, boron, barium, mercury, and hydrogen with reference to the cross-section of cadmium, in aqueous solutions of suitable salts.[44] Such measurements were soon outdated by work in Atomic Energy Projects.

When Franco Rasetti came to Université Laval in 1939 as professor and head of the new Department of Physics, he initiated a program of research on resonance and thermal capture of neutrons. His collaborators were C. Lapointe and H. Feeny.[45] In Rome Rasetti had been a member of Fermi's group, whose extensive work on the capture of neutrons in $(n, \gamma), (n, p)$, and (n, α) reactions is well known. Rasetti's choice of program at Laval was therefore natural. University work on neutrons was soon overtaken by secrecy and abandoned.

In the 1930s nuclear physics was not regarded as a legitimate activity of the National Research Council of Canada. Of the several sections of the laboratory organization, the Radium and X-ray Section had interests most closely related to nuclear physics. Returning to Canada in 1930, G. C. Laurence was appointed head of that section. The priorities of the new laboratory of the Council were those of a Bureau of Standards. Laurence had to set up standards and standardization procedures concerning radium and X rays for radiation therapy and industrial radiography. The standardization for certification of radium sources, extracted from the rich pitchblende deposit on the shore of Great Bear Lake, became a continuing service by Laurence's section. He also contributed to the theory of the thimble ionization chamber as a dosimeter, and wrote technical bulletins to assist radiologists in making better quantitative measurements in their hospital departments.

A nuclear-physical investigation became necessary when the claim was made that Canadian radium contained a significant admixture of mesothorium 1. This isotope of radium was one of Otto Hahn's discoveries. Since the half-lives of radium and mesothorium 1 are 1600 and 5.8 years respectively, a mixed source for gamma-ray therapy would deteriorate rapidly in value. Laurence was able to prove that the ore deposits at LaBine Point are practically free from mesothorium 1.[46]

Accelerator physics began in the 1930s with the construction and operation of a voltage-multiplier by J. D. Cockcroft and E. T. S. Walton at the University of Cambridge and of a cyclotron by E. O. Lawrence and M. S. Livingston at the University of California. The importance of these inventions for promoting nuclear reactions with artificially accelerated protons and other particles was immediately grasped by Professor J. A. Gray at Queen's University and Professor J. S. Foster at McGill University, and they began to make plans.

Queen's University was at first unable to help financially with Gray's plans, since the years 1933–1935 coincided with the greatest depth of the Depression, but he was encouraged to solicit donations in Canada and England. He failed to obtain promises of the required amount of money before the expiry of a favorable quotation on a 600-kilovolt voltage-multiplier. In 1937–1939 he tried again, with first choice a 37-inch cyclotron at an estimated cost of $22,000. The Board of Trustees of Queen's University could not provide this amount, but by the summer of 1939 they had given full approval for a Van de Graaff electrostatic generator at a cost of $7000. Within days World War II broke out, and by mutual agreement the plans were postponed indefinitely.[47]

From about the year 1935 Foster had been thinking about a cyclotron and a radiation laboratory at McGill. In planning, it was helpful that E. O. Lawrence and Foster had been friends and classmates at Yale University. In the autumn of 1937 the Governors of McGill University voted to provide the necessary funds. The war broke out just as the plans were complete, and the university authorities considered it unwise to proceed with the project.[48]

When the Canadian Atomic Energy Project was started during World War II there were no accelerators in Canada for making measurements on neutrons toward the design of a natural uranium–heavy water reactor as a pilot plant for the production of plutonium for atomic bombs. If either Gray or Foster had succeeded in establishing an accelerator laboratory for nuclear physics before the outbreak of war, its facilities would have been extensively used in the Atomic Energy Project, and the course of the history of science in Canada would have been greatly altered.

Queen's University

NOTES

1 Wilfrid Eggleston, *National Research in Canada, The NRC, 1916–1966* (Toronto: Clarke, Irwin, 1978), Ch. 4.
2 E. Rutherford, *Radioactive Substances and their Radiations* (Cambridge: Cambridge University Press, 1913), p. 256.
3 James Chadwick, *Verhandl. Deut. Phys. Ges.,* 16 (1914), 383.
4 James Chadwick, "Some Personal Notes on the Search for the Neutron," *Proceedings of the 10th International Congress of the History of Science,* Ithaca, 1962 (Paris: Hermann, 1964), pp. 159–162.
5 Jean Becquerel, *La découverte de la radioactivité,* Conférences prononcées à l'occasion du Cinquantième Anniversaire de la Découverte de la Radioactivité (Paris, October 1946).
6 Of Professor Gray's former research students, H. M. Cave and the writer returned to Canada in 1930 to take up positions as lecturers in physics at Queen's University. W. J. Henderson, upon his return from the Cavendish Laboratory in 1935, had to be supported for a few months in Gray's laboratory until a position could be found elsewhere. J. S. Marshall divided his detailed investigation of the beta spectra of uranium X_1 and uranium X_2 between the Cavendish Laboratory and Gray's laboratory. J. S. Marshall, *Proc. Roy. Soc. Lond.,* A173 (1939), 391.
7 This book, bound in red leather and lettered in gold, is inscribed "The Professor Wilson Prize awarded to Joseph Alexander Gray, (signed) Registrar, 10 May 1905, The University of Melbourne." It was given by Professor Gray to the Synchrotron Laboratory of Queen's University at the time of his retirement, October 1952.
8 L. A. DuBridge and H. Brown, *Rev. Sci. Instrum.,* 4 (1933), 532.
9 H. Le Caine and J. H. Waghorne, *Can J. Res.,* A19 (1941), 21.
10 J. S. Marshall and A. G. Ward, *Can J. Res.,* A15 (1937), 39.
11 B. W. Sargent, *Can. J. Res.,* A17 (1939), 82.
12 A. von Grosse, *Proc. Roy. Soc. Lond.,* A150 (1935), 363.
13 B. W. Sargent, *Proc. Roy. Soc. Lond.,* A139 (1933), 659.
14 *Table of Isotopes,* C. Michael Lederer and Virginia S. Shirley (eds.), (7th ed., New York: Wiley, 1978).
15 The degrees of "forbidenness" are now more numerous. See Emilio Segrè, *Nuclei and Particles* (New York: Benjamin, 1964), Ch. 9.
16 B. W. Sargent, *Proc. Camb. Phil. Soc.,* 28 (1932), 538.
17 E. Fermi, *Z. Phys.,* 88 (1934), 161.
18 C. D. Ellis and N. F. Mott, *Proc. Roy. Soc. Lond.,* A141 (1933), 502.
19 W. J. Henderson, *Proc. Roy. Soc. Lond.,* A147 (1934), 572.
20 The particle-antiparticle distinction by Dirac has led to the neutrino being renamed the anti-neutrino. The old name will be retained throughout this discussion.
21 Wolfgang Pauli, *Aufsätze und Vorträge über Physik und Erkenntnistheorie* (Braunschweig: Vieweg, 1961). George Gamow, *Structure of Atomic Nuclei and Nuclear Transformations* (Oxford: Clarendon Press, 1937), Ch. 7. Segrè, *Nuclei and Particles,* Ch. 9.
22 C. D. Ellis and W. A. Wooster, *Proc. Roy. Soc. Lond.,* A117 (1927), 109. L. Meitner and W. Orthmann, *Z. Phys.,* 60 (1930), 143.
23 Ernest Rutherford, James Chadwick, and C. D. Ellis, *Radiations from Radioactive Substances* (Cambridge: Cambridge University Press, 1930), pp. 392–397.

[24] "Proceedings of the American Physical Society," *Phys. Rev.*, 38 (1931), 579.

[25] Pauli, *Physik und Erkenntnistheorie*.

[26] "Structure et propriétés des noyaux atomiques," Rapports et Discussions, *Septième Conseil de Physique, L'Institut International de Physique Solvay* (Paris: Gauthier-Villars, 1934).

[27] *Z. Phys.*, 88 (1934), 161.

[28] C. D. Ellis, "The β-ray Type of Radioactive Disintegration," *International Conference on Physics*, London, 1934. Vol. I: *Nuclear Physics* (Cambridge: Cambridge University Press, 1935), pp. 52–54.

[29] N. Feather, *Proc. Camb. Phil. Soc.*, 34 (1938), 599. B. W. Sargent, *Phys. Rev.*, 54 (1938), 232, and *Can. J. Res.*, A17 (1939), 82.

[30] A. G. Ward and J. A. Gray, *Can. J. Res.*, A15 (1937), 42.

[31] Lederer and Shirley, eds., *Table of Isotopes*.

[32] B. L. Moore, Ph.D. Thesis, Cornell University (1940); also *Phys. Rev.*, 57 (1940), 355.

[33] W. B. Lewis, "George Hugh Henderson, 1892–1949," *Obit. Not. FRS*, 7 (1950), 155.

[34] G. C. Laurence, *Phil. Mag.*, 5 (1928), 1027. G. H. Henderson and J. L. Nickerson, *Phys. Rev.*, 36 (1930), 1344.

[35] G. H. Henderson and S. Bateson, *Proc. Roy. Soc. Lond.*, A145 (1934), 563. G. H. Henderson and L. G. Turnbull, *ibid.*, 582. G. H. Henderson, *ibid.*, 591. G. H. Henderson, C. M. Mushkat, and D. P. Crawford, *ibid.*, A158 (1937), 199. G. H. Henderson and F. W. Sparks, *ibid.*, A173 (1939), 238. G. H. Henderson, *ibid.*, 250.

[36] Henderson and Sparks, *Proc. Roy. Soc. Lond*, A173 (1939), 238.

[37] Henderson and Bateson, *Proc. Roy. Soc. Lond.*, A145 (1934), 563.

[38] Henderson, *Proc. Roy. Soc. Lond.*, A145 (1934), 591.

[39] Henderson, *Proc. Roy. Soc. Lond.*, A173 (1939), 250.

[40] Robert V. Gentry, "Radioactive Halos," *Annu. Rev. Nuc. Sci.* (E. Segrè, ed.), 23 (1973), 347.

[41] R. L. Fleischer, P. B. Price, and R. M. Walker, *Science*, 149 (1965), 383.

[42] B. W. Currie, "Ertle Leslie Harrington, 1887–1956," *Biogr. Sketches of Deceased Members, Proc. Roy. Soc. Can.*, 50, Ser. III (1956), 91.

[43] E. L. Harrington and O. A. Gratias, *Phil. Mag.*, 11 (1931), 285. E. L. Harrington and E. O. Braaten, *Trans. Roy. Soc. Can.*, 26, Ser. III (1932), 177.

[44] E. L. Harrington and J. L. Stewart, *Can J. Res.*, A19 (1941), 33.

[45] C. Lapointe and F. Rasetti, *Phys. Rev.*, 58 (1940), 554. F. Rasetti, *ibid.*, 869. H. Feeny, C. Lapointe, and F. Rasetti, *ibid.*, 61 (1942), 469.

[46] G. C. Laurence, *Radium Dosage*, Bull. No. 17 (Ottawa: National Research Council of Canada, 1936), p. 15.

[47] B. W. Sargent, "Joseph Alexander Gray, 1884–1966," *Biogr. Sketches of Deceased Members, Proc. Roy. Soc. Can.*, 6, Ser. IV (1968), 107.

[48] J. S. Foster, Presidential Address to Section III, Royal Society of Canada, *Trans. Roy. Soc. Can.*, 43, Ser. III (1949), 12. R. E. Bell, "John Stuart Foster, 1890–1964," *Biogr. Mem. FRS*, 12 (1966), 147.

INDEX

241

THE UNIVERSITY OF WESTERN ONTARIO
SERIES IN PHILOSOPHY OF SCIENCE

A Series of Books in Philosophy of Science, Methodology, Epistemology, Logic, History of Science, and Related Fields

8. J. M. Nicholas (ed.), *Images, Perception, and Knowledge.* Papers deriving from and related to the Philosophy of Science Workshop at Ontario, Canada, May 1974. 1977, ix+309 pp.
9. R. E. Butts and J. Hintikka (eds.), *Logic, Foundations of Mathematics, and Computability Theory.* Part One of the Proceedings of the Fifth International Congress of Logic, Methodology and Philosophy of Science, London, Ontario, Canada, 1975. 1977, x+406 pp.
10. R. E. Butts and J. Hintikka (eds.), *Foundational Problems in the Special Sciences.* Part Two of the Proceedings of the Fifth International Congress of Logic, Methodology and Philosophy of Science, London, Ontario, Canada, 1975. 1977, x+427 pp.
11. R. E. Butts and J. Hintikka (eds.), *Basic Problems in Methodology and Linguistics.* Part Three of the Proceedings of the Fifth International Congress of Logic, Methodology and Philosophy of Science, London, Ontario, Canada, 1975. 1977, x+321 pp.
12. R. E. Butts and J. Hintikka (eds.), *Historical and Philosophical Dimensions of Logic, Methodology and Philosophy of Science.* Part Four of the Proceedings of the Fifth International Congress of Logic, Methodology and Philosophy of Science, London, Ontario, Canada, 1975. 1977, x+336 pp.
13. C. A. Hooker (ed.), *Foundations and Applications of Decision Theory,* 2 volumes. Vol. I: *Theoretical Foundations.* 1978, xxiii+442 pp. Vol. II: *Epistemic and Social Applications.* 1978, xxiii+206 pp.
14. R. E. Butts and J. C. Pitt (eds.), *New Perspectives on Galileo.* Papers deriving from and related to a workshop on Galileo held at Virginia Polutechnic Institute and State University, 1975. 1978, xvi+262 pp.
15. W. L. Harper, R. Stalnaker, and G. Pearce (eds.), *Ifs. Conditionals, Belief, Decision, Chance, and Time.* 1980, ix+345 pp.
16. J. C. Pitt (ed.), *Philosophy in Economics.* Papers deriving from and related to a workshop on Testability and Explanation in Economics held at Virginia Poly-Technic Institute and State University, 1979. 1981.
17. Michael Ruse, *Is Science Sexist?* 1981, xix+299 pp.
18. Nicholas Rescher, *Leibniz's Metaphysics of Nature.* 1981, xiv+126 pp.
19. Larry Laudan, *Science and Hypothesis.* 1981, x+258 pp.
20. William R. Shea, *Nature Mathematized.* Papers deriving from the third international Conference on the history and philosophy of science, Montreal, Canada, 1981. Vol. I, 1983.
21. Michael Ruse, *Nature Animated.* Papers deriving from the third international Conference on the history and philosophy of science, Montreal, Canada, 1981. Vol. II, 1983.